Cellular Communication
in Plants

Cellular Communication
in Plants

Edited by
Richard M. Amasino
University of Wisconsin–Madison
Madison, Wisconsin

PLENUM PRESS • NEW YORK AND LONDON

Library of Congress Cataloging-in-Publication Data

Cellular communication in plants / edited by Richard M. Amasino.
      p.   cm.
   "Proceedings of the Twenty-first Steenbock Symposium: Cellular
Communication in Plants, held May 31-June 2, 1992, in Madison,
Wisconsin"--T.p. verso.
   Includes bibliographical references and indexes.
   ISBN 0-306-44415-1
   1. Plant cell interaction--Congresses.   I. Amasino, Richard M.,
 1956-   . II. Steenbock Symposium (21st : 1992 : Madison, Wis.)
QK725.C392 1993
581.87--dc20   93-10041
                                                               CIP

Proceedings of the Twenty-First Steenbock Symposium: Cellular
Communication in Plants, held May 31–June 2, 1992, in Madison, Wisconsin

ISBN 0-306-44415-1

© 1993 Plenum Press, New York
A Division of Plenum Publishing Corporation
233 Spring Street, New York, N.Y. 10013

All rights reserved

No part of this book may be reproduced, stored in a retrieval system, or
transmitted in any form or by any means, electronic, mechanical,
photocopying, microfilming, recording, or otherwise, without written
permission from the Publisher

Printed in the United States of America

Harry Steenbock (1886–1967), distinguished Professor of Biochemistry at the University of Wisconsin–Madison, achieved fame for his discovery of the irradiation process for producing vitamin D. However, his interests and contributions ranged over many areas of nutrition and biochemistry. In recognition of his contributions to biochemistry in general and the University of Wisconsin in particular, a symposium has been established. This program represents the Twenty-First Steenbock Symposium.

SYMPOSIUM COORDINATORS

Richard M. Amasino
Department of Biochemistry
University of Wisconsin-Madison
420 Henry Mall
Madison, Wisconsin 53706

Anthony B. Bleecker
Department of Botany
University of Wisconsin-Madison
B129 Birge Hall
Madison, Wisconsin 53706

Michael R. Sussman
Department of Horticulture
University of Wisconsin-Madison
395 Horticulture
Madison, Wisconsin 53706

The Steenbock Symposia are supported by an endowment from the Wisconsin Alumni Research Foundation with funds provided by Professor Harry Steenbock.

Contributions to the Twenty-first Steenbock Symposium have also been provided by:

Amoco Corporation
Monsanto Company
Pioneer Hi-Bred

PREFACE

This is an exciting period in plant biology as many disciplines, such as genetics and biochemistry, are merging to provide a more detailed understanding of plant growth and development. The purpose of this meeting was to provide a sampling of some of this exciting work in the area of cellular communication and signal transduction.

R.M. Amasino

CONTENTS

Auxin-Binding Proteins and their Possible Role in
       Cell Development 1
  Alan M. Jones

Signal Perception in Plants: Hepta-B-Glucoside Elicitor
       Binding Proteins in Soybean 7
  *Michael G. Hahn, Jong-Joo Cheong, Robert M. Alba,
    and François Côte*

The Role of Salicylic Acid as a Plant Signal Molecule 15
  Paul Silverman, Rebecca A. Linzer, and Ilya Raskin

Blue-Light Regulated Gene Expression 21
  *Lon S. Kaufman, Kathleen A. Marrs, Katherine M.F. Warpeha,
    Jie Gao, Keshab Bhattacharya, Judi Tilghman, and
    John F. Marsh III*

Role of the Maize *Viviparous-1* Gene in Regulation of Seed Maturation 27
  Donald R. McCarty

Lovastatin Induces Cytokinin Dependence in Tobacco Cultures 37
  Dring N. Crowell and Michael S. Salaz

Molecular Genetic Approaches to Elucidating the Role of
       Hormones in Plant Development 45
  Harry Klee and Charles Romano

Reversible Inhibition of Tomato Fruit Ripening by Antisense
       ACC Synthase RNA 51
  Athanasios Theologis, Paul W. Oeller and Lu Min-Wong

Genetic Dissection of Signal Transduction Pathways that
       Regulate *cab* Gene Expression 57
  *Joanne Chory, Lothar Altschmied, Hector Cabrera,
    Hsou-min Li, and Ronald Susek*

The Role of *KN1* in Plant Development 63
  Sarah Hake

Genetic Analysis of Meristem Structure and Function in
    Arabidopsis thaliana 69
  M. Kathryn Barton

Axillary Bud Development in Pea: Apical Dominance, Growth Cycles,
    Hormonal Regulation and Plant Architecture 75
  Joel P. Stafstrom

Elicitation of Organized Pigmentation Patterns by a Chalcone
    Synthase Transgene 87
  Richard A. Jorgenson

Pattern Formation During *Arabidopsis* Embryo Development 93
  Ulrike Mayer, Thomas Berleth, Ramon A. Torres Ruiz,
    Simon Miséra, and Gerd Jürgens

Cell Communication and the Coordination of Differentiation 99
  Judith A. Verbeke

Structure and Expression of Style-Expressed and Pollen-Expressed
    Components of Gametophytic Self-Incompatibility
    in *Petunia hybrida* 105
  T.L. Sims, J.J. Okuley, K.R. Clark, and P.D. Collins

Leafy Controls Meristem Identity in *Arabidopsis* 115
  Detlef Weigel and Elliot M. Meyerowitz

Arabidopsis as a Model System for Analysis of Leaf Senescence
    and Inflorescence-Meristem Longevity 123
  Linda L. Hensel and Anthony B. Bleecker

Analysis of a Receptor-Like Protein Kinase of *Arabidopsis thaliana* 131
  G. Eric Schaller, Sara Patterson, and Anthony B. Bleecker

Abstracts of Talks 139

Abstracts of Short Talks 149

Abstracts of Posters 155

Participants 169

Author Index 177

Subject Index 179

AUXIN-BINDING PROTEINS AND THEIR POSSIBLE ROLE IN CELL DEVELOPMENT

Alan M. Jones

Department of Biology
University of North Carolina
Chapel Hill, NC 27599-3280

INTRODUCTION

The standard approach toward understanding cell communication has been to utilize genetic or biochemical methods to first identify elements in signal pathways then to use biochemical methods to establish the physical relationships between these elements. Typically, a signal-regulated response such as gene expression is identified or a putative receptor is identified and then from these respective positions one "walks back" to the receptor or "walks forward" to the response. For the type of communication imparted by auxins in plant cells, *both* putative receptors and rapid responses have been identified. The walk has started and the question is whether or not the respective research groups will soon meet in the middle, having completed a signal transduction chain.

One reason for uncertainty is that there are several candidates for receptors and several rapid auxin-induced responses that could play roles in several auxin-mediated changes in cell development. Will we eventually find that there are multiple independent signal transduction pathways for multiple auxin responses or that some or all auxin responses utilize multiple receptors, multiple inputs?

AUXIN-INDUCED CELL ELONGATION: MULTIPLE EFFECTS

Primary actions and targets of auxin have been sought by studies on very rapid auxin effects on auxin-depleted tissues and as expected, these effects are many and diverse. Auxin alters calcium flux (Gehring, et al, 1990) and cytoplasmic streaming (Sweeney and Thimann, 1942) within seconds and enzyme activities (e.g. Ray, 1985; Mohnen, et al., 1985) and gene expression (Key, 1989) within minutes. On this same timescale, the cell wall becomes loosened and the cell expands, driven by turgor pressure which changes very little over time. Expansion occurs at a rate of typically 10μm/hour for hours resulting in a 5%/hour increase in cell volume. Studies have narrowed the possibilities for a primary effect of changes in cell wall property and synthesis. Most notable of these changes is an increase in acidification of the cell wall space (Senn and Goldsmith, 1988) and membrane hyperpolarization brought about by outwardly directed proton pumps (H^+ATPase). New cell wall materials are added via membrane localized synthases (Staehelin and Giddings, 1982) and through exocytotic delivery (Schnepf, 1986). Not only are the structural components most likely added through exocytosis but also enzymes such as the H^+ATPase (Hager, et al., 1991). Expansion, cell wall synthesis and proton secretion are rapidly inhibited by cyclohexamide (Edelmann and Schopfer, 1989) indicating that this complex process is controlled by growth limiting proteins with short half lives (Vanderhoef, et al., 1976). It is also known that cell expansion and cell wall synthesis are not coupled obligatorily (Bret-Harte, et al., 1991).

There is little controversy over the timing and nature of these auxin effects but what can not be agreed upon is the primary rate limiting site of action, the component(s) most directly affected by auxin. Discussion during the 1980's took two views: auxin effects on genic activity vs. auxin action at the cell wall is the direct cause of growth and there was not (and still not) agreement about whether auxin operated cytoplasmically or extracytoplasmically. This is because the timing of cellular and extracellular auxin effects were coincident, and limiting auxin to the outside (Venis, et al., 1990) or inside (Vesper and Kuss, 1990) of the cell did not induce maximal growth rates. As a result, it seemed in retrospect that the researchers working on cell wall effects discounted auxin action at gene regulation and that those working on auxin induced gene expression felt that the cell wall changes were brought about by their particular gene product. I would like to present a different view, one more complicated but probably closer to reality.

MULTIPLE AUXIN-BINDING PROTEINS

The development of azidoindole-3-acetic acid as a photoaffinity labeling reagent has recently resulted in the identity of several different auxin-binding proteins (ABP). The first ABP's identified by photoaffinity labeling were shown to have subunit sizes of 22, 24, and 43 kDa. Of these three, the most is known about the 22-kDa subunit protein, now designated ABP1. ABP1 is an endoplasmic reticulum (ER) protein by two criteria: one is cofractionation of ABP1 with ER markers (Jones, et al., 1989), and the other is the presence of an ER retention signal sequence (Inohara, et al., 1989) but there is some evidence that ABP1 may act directly at the plasma membrane. Antibodies to ABP1 block auxin induced hyperpolarization of tobacco protoplasts suggesting that ABP1 is accessible to the antiABP1 antibodies on the protoplast surface (Barbier-Brygoo, et al., 1991). Recent immunolocalization of ABP1 using electron microscopy demonstrates that some ABP1 is indeed located at the plasma membrane. These results are supported by parallel secretion studies that show specific movement of ABP1 out of the cell. Furthermore, auxin added to the medium of these cells will modulate the relative distribution of ABP causing more ABP to be retained in the cell vs. being secreted (Jones & Herman, unpublished).

Little is known of the 24-kDa subunit ABP identified by Jones and Venis (1989) although recently Feldwisch, et al (1992) confirmed the previous photolabeling described by Jones and Venis and have added that this ABP is probably located at the plasma membrane as well.

Hicks, et al (1989a and 1989b) have used photoaffinity labeling to identify a 42-kDa subunit ABP found on the plasma membrane from zucchini and tomato. This may be the dicot homolog of the 43-kDa subunit ABP identified by Jones and Venis (1989). The function of this ABP is not known, however Hicks, et al (1989a) have suggested that it may be the auxin uptake carrier based upon the similarity of a series of auxins for uptake efficiency compared to the efficiency of blocking photolabeling of the 43-kDa ABP by azidoindole-3-acetic acid.

Macdonald, et al (1991) have also used photoaffinity labeling to identify soluble ABPs in *Hyoscyamus muticus*. A 25-kDa ABP was photolabeled at 196°C and displayed a narrow specificity toward auxins and a sharp pH optimum for labeling while a 31-kDa protein was photolabeled at 4°C over a wider pH range and displayed a broader specificity for indoles. Macdonald, et al (1991) demonstrated that the 31-kDa protein was β1-3 glucanse and as yet it is uncertain if photolabeling was artifactual or if β1-3 glucanase is a genuine ABP whose activity may be modulated by auxin. It is known that β1-3 glucanase levels are indirectly regulated by auxin (Mohnen, et al., 1985) but the significance of this observation to the photolabeling is unknown.

Prasad and Jones (1991) have used anti-idiotypic antibodies to identify a 65-kDa subunit protein that appears to be a new ABP. The anti-idiotypic antibodies recognized the same 65 kDa molecular weight band in a variety of plants and it was shown that antibody recognition in Western blots could be blocked by auxins. Furthermore, the anti-idiotypic antibodies retained auxin binding activity of soybean extract on an immunoaffinity. Finally, pure nuclei were shown to contain this ABP. Partial purification on DNA cellulose suggests that this ABP binds DNA, consistent with our working hypothesis that this ABP is a ligand-regulated transcriptional enhancer such as has been shown for the members of the family of steroid receptors.

MULTIPLE EFFECTS: MULTIPLE OR SINGLE RECEPTORS?

As yet, there is no evidence to link the activity of any of the ABPs described above to any of the auxin-induced events of cell elongation . In other words, we can not yet say that any of the ABP's have receptor function. Nor do we know how many receptors are expected to be involved in auxin-induced cell elongation or if different auxin responses are individually mediated by separate receptors. However, at this point since there are several candidates for auxin receptors we must consider the possibility of multiple receptors mediating a single response. Obviously, some of the ABP's will be involved in auxin metabolism and auxin movement therefore we need not assume that *all* the discovered ABP's are directly mediating cell growth.

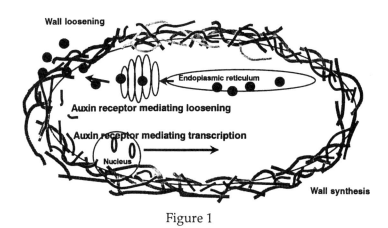

Figure 1

I conclude with the following hypothesis summarized in figure 1. Auxin induces cells to elongate by changing the cell wall properties thus allowing turgor to expand the cell. This effect, called wall loosening, is mediated by the 22-kDa ABP or related ABP and the movement of ABP from ER to plasma membrane discussed above may be important to this role for ABP1. Cell wall expansion needs rapid addition of new cell materials which possibly involves regulation of specific gene expression. The 65-kDa ABP regulates specific gene expression to ultimately provide these new wall materials. The movement in or out of the cell is mediated by plasma membrane located auxin transporters and the 40-kDa ABP may be the uptake carrier for auxin. Finally, cytoplasmic ABP are expected. These ABP's (not shown) would fulfill various functions such as IAA synthesis, degradation and conjugation.

This hypothesis is highly speculative, but testable. Given the rate of progress we are now making, it is my view that in the next few years we will be able to replace the above hypothesis with one that is more accurate.

REFERENCES

Barbier-Brygoo, et al., 1991, Perception of the auxin signal at the plasma membrane of tobacco mesophyll protoplasts. *Plant Journal* 1:83.

Bret-Harte, M.S., Baskin, T.I., Green, P.B., 1991, Auxin stimulates both deposition and breakdown of materials in the pea outer epidermal cell wall, as measured interferometrically, *Planta* 185:462.

Edelmann, H., Schopfer, P., 1989, Role of protein and RNA synthesis in the initiation of auxin-mediated growth in coleoptiles of coleoptiles of Zea mays. L., Planta 179:475.

Feldwisch, et al., 1992, An auxin-binding protein is localized to the plasma membrane of maize coleoptile cells: Identification by photoaffinity labeling and purification of a 23-kDa polypeptide. *Proc. Natl. Acad. Sci. USA* 89:475.

Gehring, C.A., Irving, H.R., Parish, R.W., 1990, Effects of auxin and abscisic acid on cytosolic calcium and pH in plant cells. *Proc. Natl. Acad. Sci. USA* 87: 9645.

Hager, A., Debus, G., Edel, H.-G., Stransky, H., Serrano, R., 1991, Auxin induces exocytosis and the rapid synthesis of a high-turnover pool of plasma-membrane H+ATPase, *Planta* 185:527.

Hicks, G.R., Rayle, D.L., Jones, A.M., Lomax, T.L., 1989a, Specific photoaffinity labeling of two plasma membrane polypeptides with an azido auxin. *Proc. Natl. Acad. Sci. USA* 86:4948.

Hicks, G.R., Rayle, D.L., Lomax, T.L., 1989b, The *Diageotropica* mutant of tomato lacks high specific activity auxin binding sites. *Science* 245:52.

Inohara, N., Shimomura, S., Fukui, T., Futai, M., 1989, Auxin-binding protein located in the endoplasmic reticulum of maize shoots: molecular cloning and complete primary structure. *Proc. Natl. Acad. Sci. USA* 86: 3564.

Jones, A.M., Venis, M.A., 1989, Photoaffinity labeling of indole-3-acetic acid-binding proteins in maize, *Proc. Natl. Acad. Sci. USA* 86: 6153.

Jones, A.M., Lamerson, P., Venis, M.A., 1989, Comparison of site I auxin binding and a 22-kDa auxin-binding protein in maize. *Planta* 179:409.

Key, J.L., 1989, Modulation of gene expression by auxin, *BioEssays* 11:52.

Macdonald, H., Jones, A.M., King, P.J., 1991, Photoaffinity labeling of soluble auxin-binding proteins. *J. Biol. Chem.* 266:7393.

Mohnen, D., Shinshi, H., Felix, F., and Meins, F., Jr., 1985, Hormonal regulation of b1,3-glucanase messenger RNA levels in cultured tobacco tissues, *EMBO J.* 4:1631.

Prasad, P.V., Jones, A.M., 1991 Putative receptor for the plant growth hormone auxin identified and characterized by anti-idiotypic antibodies. *Proc. Natl. Acad. Sci. USA* 88:5479.

Ray, P.M., 1985, Auxin and fusicoccin enhancement of b-glucan synthase in peas. *Plant Physiol.* 78:466.

Schnepf, E., 1986, Cellular polarity, *Ann. Rev. Plant Physiol.* 37:23.

Senn, A.P., Goldsmith, M.H., 1988 Regulation of electrogenic proton pumping by auxin and fusicoccin as related to the growth of *Avena* coleoptiles. *Plant Physiol.* 88:131.

Staehlin, L.A., Giddings, T.H., 1982 Membrane-mediated control of cell wall microfibrillar order. in: "Developmental order: its origin and regulation" S. Subtelny and p. B. Green, eds. Alan Liss, New York, p133.

Sweeney, B.M., Thimann, K.V., 1942, The effects of auxin on protoplasmic streaming, *J. Gen. Physiol.* 25:841.

Vanderhoef, L.N., Stahl, C.A., Lu, T.Y-S., 1976, Two elongation responses to auxin respond differently to protein synthesis inhibition. *Plant Physiol.* 58:402.

Venis, M.A., et al., 1990, Impermeant auxin analogues have auxin activity, *Planta* 182:232.

Vesper, M.J., Kuss, C.L., 1990, Physiological evidence that the primary site of auxin action in maize coleoptiles is an intracellular site. *Planta* 182:486.

SIGNAL PERCEPTION IN PLANTS: HEPTA-β-GLUCOSIDE ELICITOR BINDING PROTEINS IN SOYBEAN

Michael G. Hahn, Jong-Joo Cheong, Robert M. Alba, and François Côté

Complex Carbohydrate Research Center
The University of Georgia
220 Riverbend Road
Athens, GA 30602-4712
USA

INTRODUCTION

Living organisms utilize a large number of signal molecules to regulate their growth and development. Furthermore, the cells that make up an organism have evolved complex and diverse mechanisms for perceiving and responding to signal molecules originating not only from within the organism, but also from the external environment. Biochemical analysis of the interactions between plants and microbes has contributed to a greater understanding of the molecular basis for signal perception and transduction in plant cells (for recent reviews, see [1-3]). Research in this area has led to the discovery of new classes of signal molecules and provided useful model systems for molecular studies on signal perception, signal transduction, and gene regulation in plants. This article will give an overview of our studies of one plant signal transduction system, the induction of phytoalexin accumulation by oligoglucoside elicitors. Our research to date has focussed on the first stage of this signal transduction system, the specific recognition by plant receptors of molecules (elicitors) that induce phytoalexin accumulation.

GLUCAN ELICITORS: STRUCTURE-ACTIVITY STUDIES

The synthesis in plants of antimicrobial phytoalexins is one of the best studied plant defense responses induced upon infection with a variety of microorganisms [1,4]. Phytoalexin synthesis and accumulation are observed not only after microbial infection, but also after treatment of plant tissues with elicitors. The term "elicitor," originally used to refer to molecules and other stimuli that induce phytoalexin synthesis in plant cells [5], is now commonly used for molecules that stimulate any plant defense mechanism [1,4,6]. A number of different cell wall constituents, originating either from the host plant or from the invading microbe, can induce phytoalexin accumulation and other plant defense responses [7]. Among these elicitors, the elicitors that are derived from glucans present in the mycelial walls of a number of fungi are the best characterized.

Elicitor-active oligosaccharides were first detected in the culture medium [8] and then in the mycelial walls [9,10] of *Phytophthora megasperma* f. sp. *glycinea*, a fungal pathogen of soybean. A hepta-β-glucoside (Figure 1), purified from the mixture of oligoglucosides generated by partial acid hydrolysis of *P. megasperma* mycelial walls, was the smallest oligoglucoside that induces soybean seedlings to accumulate phytoalexins [11]. The bioactive hepta-β-glucoside was purified from a mixture estimated to contain ~300 structurally distinct, elicitor-inactive hepta-β-glucosides. Homogeneous preparations of the aldehyde-reduced forms (i.e., the hexa-β-glucosyl glucitols) of the elicitor-active hepta-β-glucoside and of seven other elicitor-inactive hepta-β-glucosides were obtained in amounts sufficient to determine their structures [11,12]. The structure of the elicitor-active hepta-β-glucoside [12] has been confirmed by chemical synthesis [13-15].

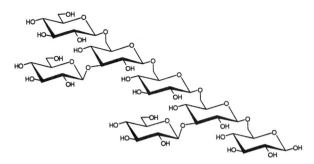

Figure 1. Structure of the hepta-β-glucoside elicitor.

The ability of the chemically synthesized, unreduced hepta-β-glucoside elicitor to induce phytoalexin accumulation in soybean cotyledons is identical to that of the corresponding hexa-β-glucosyl glucitol purified from fungal wall hydrolyzates [12]. They are active at concentrations of ~10 nM, making them the most active elicitors of phytoalexin accumulation yet observed. The seven other hexa-β-glucosyl glucitols that were purified from the partial hydrolyzates of fungal cell walls had no elicitor activity over the limited concentration range (≤ 400 μM) that could be tested [11]. These results provided the first evidence that specific structural features are required for an oligo-β-glucoside to be an effective elicitor of phytoalexin accumulation. Our research has built and expanded upon these initial findings.

Twelve oligo-β-glucosides (**2-13**, see Figures 2 and 3), structurally related to the elicitor-active hepta-β-glucoside (compound **1**, Figure 2) were chemically synthesized [13,14,16-18]. This allowed structural features essential for effective elicitation of phytoalexin accumulation in soybean cotyledon tissue to be identified [19]. The four most active oligo-β-glucosides [compounds **1**, **4**, **7**, and **8**; concentration required for half-maximum induction of phytoalexin accumulation (EC_{50}) ~10 nM] have the same branching pattern as that of elicitor-active hepta-β-glucoside **1** [20]. Hexa-β-glucoside **4** (Figure 2) is the minimum fully elicitor-active structure [19]. Increasing the length of this hexaglucoside by addition of glucosyl residues at the reducing end of the molecule has no significant effect on its elicitor activity. In contrast, removing glucosyl residues from the hexaglucoside (compounds **2** and **3**) or rearranging its side chains (compound **5**) results in molecules with significantly lower elicitor activity (Figure 2). For example, removal of the non-reducing terminal backbone glucosyl residue resulted in a 4,000-fold reduction in elicitor activity, suggesting that this glucosyl residue has a particularly important function. The importance of the side-chain glucosyl residues to elicitor activity was confirmed by the demonstration that a linear, 6-linked hepta-β-glucoside is inactive.

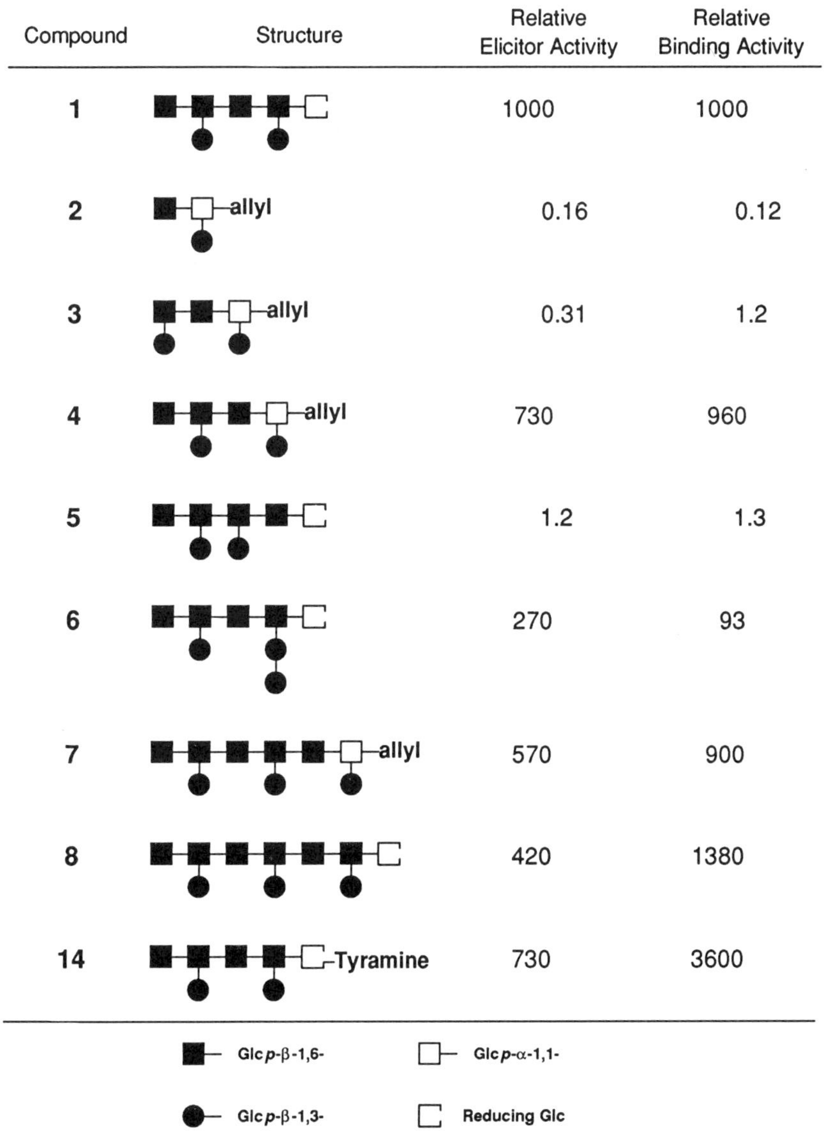

Compound	Structure	Relative Elicitor Activity	Relative Binding Activity
1		1000	1000
2	allyl	0.16	0.12
3	allyl	0.31	1.2
4	allyl	730	960
5		1.2	1.3
6		270	93
7	allyl	570	900
8		420	1380
14	Tyramine	730	3600

■— Glc$p$-β-1,6-   □— Glc$p$-α-1,1-

●— Glc$p$-β-1,3-   ▢ Reducing Glc

Figure 2. Structure, relative elicitor activity, and relative binding activity of oligo-β-glucosides. The relative elicitor activity is calculated with respect to the phytoalexin-inducing activity of hepta-β-glucoside **1** on soybean cotyledons (=1000). The relative binding activity is calculated with respect to the ability of the hepta-β-glucoside elicitor (compound **1**) to displace radiolabeled ligand (iodinated compound **14**) from soybean root membranes (=1000). Reprinted from [29] with permission.

An additional set of hexasaccharides was synthesized in which one or the other terminal glucosyl residue at the non-reducing end of hexa-β-glucoside 4 was modified (Figure 3). Thus, replacement of the side-chain glucosyl residue of the terminal trisaccharide with a β-glucosaminyl (compound 10) or N-acetyl-β-glucosaminyl residue (compound 11) reduced the elicitor activity ~10- and ~1,000-fold, respectively. The corresponding modifications of the non-reducing terminal backbone glucosyl residue (compounds 12 and 13, respectively) resulted in even greater decreases in elicitor activity (~100- and ~10,000-fold, respectively), while substituting this glucosyl residue (compound 9) with a xylosyl residue reduced the activity about 10-fold (unpublished results of the authors). These structure-activity studies confirmed that specific structural features are required for an oligo-β-glucoside to effectively elicit phytoalexin accumulation in soybean.

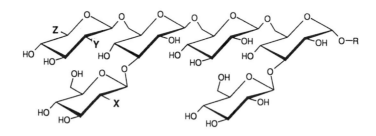

Compound	X	Y	Z	Relative Elicitor Activity	Relative Binding Activity
4	-OH	-OH	-CH$_2$OH	730	960
9	-OH	-OH	-H	130	530
10	-NH$_2$	-OH	-CH$_2$OH	130	180
11	-NHAc	-OH	-CH$_2$OH	1.2	3.8
12	-OH	-NH$_2$	-CH$_2$OH	4.4	21
13	-OH	-NHAc	-CH$_2$OH	<0.08	0.22

Figure 3. Structure, relative elicitor activity, and relative binding activity of structural variants of hexa-β-glucoside 4. The relative elicitor activity is calculated with respect to the phytoalexin-inducing activity of the hepta-β-glucoside elicitor (compound 1, see Fig. 2) on soybean cotyledons (=1000). The relative binding activity is calculated with respect to the ability of the hepta-β-glucoside elicitor (compound 1) to displace radiolabeled ligand from soybean root membranes (=1000). R = allyl for compounds 4, 9, 10, 13; R = propyl for compounds 11, 12. The nature of R at the reducing end of these hexasaccharides has little effect on elicitor or binding activities. Reprinted from [29] with permission.

Phytoalexin elicitor assays of reducing-end derivatives of hepta-β-glucoside 1 demonstrated that attachment of an alkyl or aromatic group to the oligosaccharide (e.g., compound 14) did not have a significant effect on its EC$_{50}$ [19]. A tyramine-coupled derivative of hepta-β-glucoside 5 was slightly more active (~2.5-fold) than underivatized hepta-β-glucoside 5. Coupling tyramine or benzylhydroxylamine to maltoheptaose, a structurally unrelated hepta-α-glucoside, yielded derivatives with no detectable elicitor activity. This established that coupling of aromatic groups to biologically inactive oligoglucosides does not endow those oligosaccharides with phytoalexin elicitor activity. Thus, it was possible to prepare a fully active, radio-iodinated form of compound **14** for use as a labeled ligand to search for the presence of elicitor-specific, high-affinity binding sites in soybean membranes.

GLUCAN ELICITOR BINDING PROTEINS

The first step in the signal transduction pathway induced by the hepta-β-glucoside elicitor is likely to be its recognition by a specific receptor. Indeed, the specificity of the response of soybean tissue to oligoglucoside elicitors of phytoalexin accumulation [11,19] described in the previous section suggests that a specific receptor for the hepta-β-glucoside elicitor exists in soybean cells. Several earlier studies utilizing heterogeneous mixtures of fungal glucan fragments suggested that binding sites for such fragments exist in plant membranes [21-24]. The existence of membrane-localized binding sites was demonstrated for the hepta-β-glucoside elicitor coupled to radio-iodinated aminophenethylamine [25] or tyramine [26]. Binding sites were found in membranes isolated from every major organ of young soybean plants. The elicitor binding sites co-migrated with enzyme markers (vanadate-sensitive ATPase and glucan synthase II) for plasma membranes in isopycnic sucrose density gradients (unpublished results of the authors). Binding of the radiolabeled hepta-β-glucoside elicitor to the root membranes is saturable over a concentration range of 0.1 to 5 nM, which is somewhat lower than the range of concentrations (6 to 200 nM) required to saturate the bioassay for phytoalexin accumulation [11,12,19]. The root membranes possess only a single class of high-affinity hepta-β-glucoside binding sites (apparent $K_d \approx 1$ nM), which are inactivated by heat or pronase treatment [26], suggesting that the molecule(s) responsible for the binding are proteinaceous. Binding of the active hepta-β-glucoside to the membrane preparation is reversible, indicating that the elicitor does not become covalently attached to the binding protein(s).

The membrane-localized, elicitor-binding proteins exhibit a high degree of specificity with respect to the oligoglucosides that they bind. More importantly, the ability of an oligoglucoside to bind to soybean root membranes correlates with its ability to induce phytoalexin accumulation (Figure 4; [26]). Oligo-β-glucosides with high elicitor activity are efficient competitors of the radiolabeled elicitor, while biologically less active oligo-β-glucosides are less efficient. Thus, four oligo-β-glucosides ranging in size from hexamer to decamer (compounds **1**, **4**, **7**, and **8**), indistinguishable in their abilities to induce phytoalexin accumulation [19], are equally effective competitive inhibitors of binding of radiolabeled hepta-β-glucoside **14** to soybean root membranes (Figure 2). The abilities of structurally modified oligo-β-glucosides to compete with radiolabeled **14** are reduced in proportion to the reduction in the biological activities of these oligo-β-glucosides (Figures 2 and 3).

The hepta-β-glucoside elicitor-binding proteins have been solubilized from soybean root microsomal membranes with the aid of detergents ([27,28] and unpublished results of the authors). The non-ionic detergents dodecylmaltoside, dodecylsucrose and Triton X-114 each solubilized between 40 and 60% of the elicitor-binding activity from the membranes. The zwitterionic detergent N-docecyl-N,N-dimethyl-3-ammonio-1-propane-sulfonate (ZW 3-12) was almost as effective at low detergent concentrations (~0.3%) as the non-ionic detergents, but the zwitterionic detergent inactivated a significant proportion of the elicitor-binding protein(s). The solubilized elicitor binding proteins retained their high affinity for the hepta-β-glucoside elicitor (apparent $K_d = 1.4$ nM) and, more importantly, retained the specificity for elicitor-active oligoglucosides characteristic of the membrane-localized proteins (Figure 4; [28]). The successful solubilization of the elicitor binding

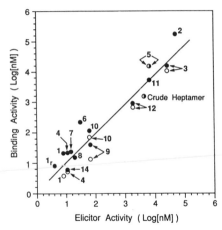

Figure 4. Correlation of the elicitor activities of oligoglucosides with their affinities for the membrane-localized (●) or detergent-solubilized (○) hepta-β-glucoside binding protein(s). The relative elicitor activity is defined as the concentration of an oligosaccharide required to give half-maximum induction of phytoalexin accumulation (A/A_{std} = 0.5) in the soybean cotyledon bioassay corrected to the standard curve for hepta-β-glucoside **1** [19]. The binding activity is defined as the concentration of oligosaccharide required to give 50% inhibition of the binding of radiolabeled hepta-β-glucoside **14** to its binding protein(s). Data points are identified with numbers identifying the oligoglucosides (see Figs. 1 and 2); 1_r = reduced hepta-β-glucoside **1**; **Crude Heptamer** = mixture of heptaglucosides prepared from mycelial wall hydrolyzates of *P. megasperma* [11]. Reprinted from [29] with permission.

proteins in a fully functional form is a crucial first step toward purification of these proteins.

CONCLUSIONS

The results of the structure-activity [19] and ligand binding [26] studies demonstrated that those structural elements of the hepta-β-glucoside required to elicit phytoalexin synthesis are also essential for efficient binding of the elicitor to its putative receptor. These essential structural features include the branched trisaccharide at the non-reducing end of the hepta-β-glucoside elicitor and the distribution of side-chain glucosyl residues along the backbone of the molecule. The combined results of the biological assays [19] and the binding studies [26] provide strong evidence that the binding protein is the physiological receptor of the hepta-β-glucoside elicitor. However, proof that this is indeed the case will require purification and characterization of elicitor binding proteins and/or their corresponding genes.

ACKNOWLEDGEMENTS

We are grateful to P. Garegg (University of Stockholm, Sweden) and T. Ogawa (RIKEN, Japan) and their colleagues for their generous gifts of synthetic oligoglucosides. We also thank C. L. Gubbins Hahn for drawing Figure 4. This work is supported by a grant from the National Science Foundation (DCB-8904574). Also supported in part by the USDA/DOE/NSF Plant Science Centers Program through funding by Department of Energy grant DE-FG09-87ER13810. François Côté is supported by a post-doctoral fellowship from the Natural Sciences and Engineering Research Council of Canada.

REFERENCES

1. R.A. Dixon, The phytoalexin response: Elicitation, signalling and control of host gene expression, *Biol. Rev.* 61:239-291 (1986).
2. C.J. Lamb, M.A. Lawton, M. Dron and R.A. Dixon, Signals and transduction mechanisms for activation of plant defenses against microbial attack, *Cell* 56:215-224 (1989).
3. D. Scheel and J.E. Parker, Elicitor recognition and signal transduction in plant defense gene activation, *Z. Naturforsch.* 45c:569-575 (1990).
4. J. Ebel, Phytoalexin synthesis: The biochemical analysis of the induction process, *Annu. Rev. Phytopathol.* 24:235-264 (1986).
5. N.T. Keen, Specific elicitors of plant phytoalexin production: Determinants of race specificity in pathogens? *Science* 187:74-75 (1975).
6. K. Hahlbrock and D. Scheel, Biochemical responses of plants to pathogens, *in:* "Innovative Approaches to Plant Disease Control," I. Chet, ed., John Wiley & Sons, Inc., New York, NY, pp. 229-254 (1987).
7. M.G. Hahn, P. Bucheli, F. Cervone, S.H. Doares, R.A. O'Neill, A. Darvill and P. Albersheim, Roles of cell wall constituents in plant-pathogen interactions, *in:* "Plant-Microbe Interactions. Molecular and Genetic Perspectives, Vol. 3," T. Kosuge and E.W. Nester, eds., McGraw Hill Publishing Co., New York, NY, pp. 131-181 (1989).
8. A.R. Ayers, J. Ebel, F. Finelli, N. Berger and P. Albersheim, Host-pathogen interactions. IX. Quantitative assays of elicitor activity and characterization of the elicitor present in the extracellular medium of cultures of *Phytophthora megasperma* var. *sojae*, *Plant Physiol.* 57:751-759 (1976).
9. A.R. Ayers, J. Ebel, B. Valent and P. Albersheim, Host pathogen interactions. X. Fractionation and biological activity of an elicitor isolated from the mycelial walls of *Phytophthora megasperma* var. *sojae*, *Plant Physiol.* 57:760-765 (1976).
10. A.R. Ayers, B. Valent, J. Ebel and P. Albersheim, Host-pathogen interactions. XI. Composition and structure of wall-released elicitor fractions, *Plant Physiol.* 57:766-774 (1976).
11. J.K. Sharp, B. Valent and P. Albersheim, Purification and partial characterization of a β-glucan fragment that elicits phytoalexin accumulation in soybean, *J. Biol. Chem.* 259:11312-11320 (1984).
12. J.K. Sharp, P. Albersheim, P. Ossowski, Å. Pilotti, P.J. Garegg and B. Lindberg, Comparison of the structures and elicitor activities of a synthetic and a mycelial-wall-derived hexa(β-D-glucopyranosyl)-D-glucitol, *J. Biol. Chem.* 259:11341-11345 (1984).
13. P. Ossowski, Å. Pilotti, P.J. Garegg and B. Lindberg, Synthesis of a glucoheptaose and a glucooctaose that elicit phytoalexin accumulation in soybean, *J. Biol. Chem.* 259:11337-11340 (1984).
14. P. Fügedi, W. Birberg, P.J. Garegg and Å. Pilotti, Syntheses of a branched heptasaccharide having phytoalexin-elicitor activity, *Carbohydr. Res.* 164:297-312 (1987).
15. J.P. Lorentzen, B. Helpap and O. Lockhoff, Synthese eines elicitoraktiven heptaglucansaccharides zur untersuchung pflanzlicher abwehrmechanismen, *Angew. Chem.* 103:1731-1732 (1991).
16. P. Fügedi, P.J. Garegg, I. Kvarnström and L. Svansson, Synthesis of a heptasaccharide, structurally related to the phytoelicitor active glucan of *Phytophthora megasperma* f.sp. *glycinea*, *J. Carbohydr. Chem.* 7:389-397 (1988).
17. W. Birberg, P. Fügedi, P.J. Garegg and Å. Pilotti, Syntheses of a heptasaccharide β-linked to an 8-methoxycarbonyl-oct-1-yl linking arm and of a decasaccharide with structures corresponding to the phytoelicitor active glucan of *Phytophthora megasperma* f.sp. *glycinea*, *J. Carbohydr. Chem.* 8:47-57 (1989).
18. N. Hong and T. Ogawa, Stereocontrolled syntheses of phytoalexin elicitor-active β-D-glucohexaoside and β-D-gluconoaoside, *Tetrahedr. Lett.* 31:3179-3182 (1990).
19. J.-J. Cheong, W. Birberg, P. Fügedi, Å. Pilotti, P.J. Garegg, N. Hong, T. Ogawa and M.G. Hahn, Structure-activity relationships of oligo-β-glucoside elicitors of phytoalexin accumulation in soybean, *Plant Cell* 3:127-136 (1991).
20. J.K. Sharp, M. McNeil and P. Albersheim, The primary structures of one elicitor-active and seven elicitor-inactive hexa(β-D-glucopyranosyl)-D-glucitols isolated from the mycelial walls of *Phytophthora megasperma* f. sp. *glycinea*, *J. Biol. Chem.* 259:11321-11336 (1984).
21. B.M. Peters, D.H. Cribbs and D.A. Stelzig, Agglutination of plant protoplasts by fungal cell wall glucans, *Science* 201:364-365 (1978).
22. M. Yoshikawa, N.T. Keen and M.-C. Wang, A receptor on soybean membranes for a fungal elicitor of phytoalexin accumulation, *Plant Physiol.* 73:497-506 (1983).

23. W.E. Schmidt and J. Ebel, Specific binding of a fungal glucan phytoalexin elicitor to membrane fractions from soybean *Glicine max*, *Proc. Natl. Acad. Sci. USA* 84:4117-4121 (1987).
24. E.G. Cosio, H. Pöpperl, W.E. Schmidt and J. Ebel, High-affinity binding of fungal β-glucan fragments to soybean (*Glycine max* L.) microsomal fractions and protoplasts, *Eur. J. Biochem.* 175:309-315 (1988).
25. E.G. Cosio, T. Frey, R. Verduyn, J. Van Boom and J. Ebel, High-affinity binding of a synthetic heptaglucoside and fungal glucan phytoalexin elicitors to soybean membranes, *FEBS Lett.* 271:223-226 (1990).
26. J.-J. Cheong and M.G. Hahn, A specific, high-affinity binding site for the hepta-β-glucoside elicitor exists in soybean membranes, *Plant Cell* 3:137-147 (1991).
27. E.G. Cosio, T. Frey and J. Ebel, Solubilization of soybean membrane binding sites for fungal β-glucans that elicit phytoalexin accumulation, *FEBS Lett.* 264:235-238 (1990).
28. J.-J. Cheong and M.G. Hahn, Solubilization and purification, from soybean root membranes, of specific binding protein(s) for a hepta-β-glucoside elicitor, *Plant Physiol.* 96(Supplement):70 (1991).
29. A. Darvill, C. Augur, C. Bergmann, R.W. Carlson, J.-J. Cheong, S. Eberhard, M.G. Hahn, V.-M. Lo, V. Marfà, B. Meyer, D. Mohnen, M.A., O'Neill, M.D. Spiro, H. van Halbeek, W.S. York and P. Albersheim, Oligosaccharins - Oligosaccharides that regulate growth, development and defense responses in plants, *Glycobiology*, in press (1992).

THE ROLE OF SALICYLIC ACID AS A PLANT SIGNAL MOLECULE

Paul Silverman, Rebecca A. Linzer, and Ilya Raskin

Center for Agricultural Molecular Biology
Cook College, P.O. Box 231
Rutgers University
New Brunswick, NJ 08903-0231

The salicylates are a widely distributed class of aromatic compounds. Modifications of the free acid result in a diverse array of compounds such as salicin, the salicyl alcohol glucoside, and methylsalicylate (oil of wintergreen). First isolated from willow (*Salix*), salicylates have been used as folk remedies since the 4th century B.C. (Weissman, 1991). Salicylates are currently used as antimicrobials, food preservatives and analgesics.

Salicylic acid (SA) is a low molecular weight compound produced in plants by phenylpropanoid metabolism. Ubiquitous in distribution, SA was found in the leaves and reproductive structures of 34 agronomically important plant species (Raskin *et al.*, 1990), and in various edible plant products (Herrmann, 1990). The observation that plants from a range of taxa contain SA suggests that this molecule may play an essential role in regulation of plant growth and development. A diverse array of effects have been observed following treatment of plants with SA (reviewed by Raskin, 1992). However, the role of SA as an endogenous regulator of these processes has remained obscure. Recently, investigations of the role of SA as an endogenous signal molecule have elucidated several functions of the molecule in plants. In this chapter, we discuss the current understanding of the role of SA as an endogenous signal in the regulation of thermogenesis and plant disease resistance.

SALICYLIC ACID AND THERMOGENESIS

As early as 1778, Lamarck described the ability of plants in the genus *Arum* to produce heat. Heat production, or thermogenicity, is now known to occur in the male reproductive structures of cycads and in the flowers or inflorescences of some angiosperm species belonging to the Annonaceae, Araceae, Aristolocheaceae, Cyclanthaceae, Nymphaeaceae and Palmae families (Meeuse and Raskin, 1988). In the arum lily, *Sauromatum guttatum* Schott, the inflorescence is surrounded by a large bract, called the spathe, which unfolds at anthesis (blooming) to expose an upper part of the spadix known as the appendix. The temperature of the appendix increases 14°C above ambient on the day of blooming (van Herk, 1937). This heat volatilizes foul-smelling indoles and amines which are attractive to insect pollinators.

In 1937, van Herk suggested that the burst of metabolic activity responsible for this temperature rise was triggered by a substance termed "calorigen". He predicted calorigen to be a water-soluble compound, produced in the male flower primordia, that diffused into the appendix on the day preceding blooming (van Herk, 1937). Despite van Herk's early theory of calorigen, the putative molecule was only recently characterized. In 1987, the search for calorigen was finally successful. Calorigen was purified and characterized from male flowers of the voodoo lily and identified as SA (Raskin et al., 1987). On the day preceding anthesis, the levels of SA in the appendix of *S. guttatum* increased almost 100-fold to 1 µg/g fresh weight. By the end of the heat production period, SA in the appendix tissue had returned to basal levels. The timing and magnitude of this temperature increase was duplicated by application of crude calorigen extract to sections of immature appendix. The application of exogenous SA at levels as low as 0.13 µg/g fresh weight lead to temperature increases of as much as 12°C. Appendix tissue treated closer to the time of anthesis had increased sensitivity to SA. Analogs of SA were also tested and only applications of 2,6 dihydroxybenzoic acid and aspirin (acetylsalicylic acid) induced thermogenesis (Raskin et al., 1989). The temperature increases were accompanied by the emission of a characteristic foul odor. These two phenomenon were inextricable.

Thermogenesis is caused by a dramatic increase in the alternative respiratory pathway, a cyanide insensitive mode of respiration found in all plants. In the alternative pathway, energy conservation sites II and III of the cytochrome c pathway are bypassed and no electrochemical gradient is created (Siedow and Berthold, 1986). Instead, most of the chemical energy of the respiratory substrates passing through the pathway are dissipated as heat. The increase in the alternative pathway respiration is so dramatic that at the peak of heat production oxygen consumption in the inflorescences of voodoo lilies is as high as that of a hummingbird in flight. (reviewed by Raskin, 1992) In order to provide substrates for this metabolic burst, activation of glycolytic and Krebs cycle enzymes is happening simultaneously with the respiratory explosion.

The voodoo lily provides an excellent system for the study of the alternative respiratory pathway. During the five days prior to anthesis, alternative oxidase activity was determined to increase over 10-fold (Elthon and McIntosh, 1986). This increase was shown to involve *de novo* synthesis of the alternative oxidase protein. The nuclear gene encoding alternative oxidase has been isolated and characterized (Hiser and McIntosh, 1990). It codes for a protein with a predicted molecular mass of 38.9 kDa (Rhoads and McIntosh, 1991). Both calorigen extract and SA induce the alternative oxidase gene, confirming the identity of calorigen to be SA.

Recently, the involvement of SA has been demonstrated in the regulation of the alternative respiratory pathway in non-overtly thermogenic plants. As plant cell suspension cultures age, they show increased capacity for alternative respiration (Horn and Mertz, 1982). Kapulnik et al. (1992) showed that the application of SA to young tobacco cell cultures will increase their heat production. The cultures become more resistant to KCN, an inhibitor of the cytochrome c pathway, and more sensitive to SHAM, an inhibitor of the alternative pathway which blocked SA-induced heat increases in the young tobacco cell cultures. The SA-induced increase in the activity and capacity of the alternative respiratory pathway was caused by a redirection of electrons from the cytochrome pathway into the alternative, heat producing, pathway. Again, 2,6 dihydroxybenzoic acid was as effective as SA in stimulating the alternative pathway, as was 4-hydroxybenzoic acid.

The discovery of the role of SA as an endogenous signal for heat production initiated investigations of other physiological processes which may be regulated by SA.

SALICYLIC ACID AND DISEASE RESISTANCE

Some plant-pathogen interactions result in a host-mediated resistance response, which restricts pathogen spread by localized necrosis. This localized necrosis is termed the hypersensitive response (HR), and may be the result of interaction with a fungal, bacterial, or viral pathogen. Following the HR, plants may become more resistant to subsequent pathogen attack - a phenomenon known as systemically acquired resistance (SAR) (Ross, 1961). SAR results in resistance to both the inducing organism or to other taxonomically distant pathogens.

The expression of at least six families of proteins is usually found in conjunction with SAR and the HR. These proteins, termed pathogenesis-related proteins (Antoniw et al., 1980), are expressed locally in association with the HR and systemically with the appearance of SAR. Although the function of the PR-1 proteins is not known, putative roles have been assigned to the PR-2 and PR-3 families as ß-1,3-glucanase and chitinase, respectively, based on sequence comparisons. SA has recently been shown to induce transcription of all six families of PR-proteins observed during SAR (Ward et al., 1991). Based on their probable functions and time of expression, PR-proteins have been postulated to play a role in acquired resistance. (reviewed by Linthorst, 1991). However, no definitive proof for PR-protein's causing SAR has been demonstrated.

The existence of a mobile chemical signal activating SAR throughout the plant was postulated more than 25 years ago (Ross, 1966, Kuc, 1982). Studies involving stem-girdling and grafting of cucumber following infection by *Colletotricum lagenarium* showed that the signal molecule in SAR is phloem-mobile (Guedes *et al.*, 1980). A candidate signal molecule in SAR would need to satisfy the following criteria: a) induce resistance to a broad suite of potential pathogens, b) induce PR-proteins, c) increase in concentration following pathogen attack, both locally and throughout the plant and d) move through the phloem.

The ability of salicylates to induce plant disease resistance was first demonstrated by White (1979). Treating TMV-resistant tobacco with aspirin (acetylsalicylic acid), which is hydrolyzed to SA, resulted in a reduction in lesion number when plants were subsequently inoculated with TMV. Furthermore, the aspirin treatments induced PR-1 proteins and TMV resistance in a dose-dependent manner. It was subsequently shown that treatment with aspirin induced resistance to TMV in TMV-susceptible tobacco varieties (White *et al.*, 1983). Applied salicylate has since been shown to induce resistance and PR-proteins in many other plant-pathogen systems (reviewed by Raskin, 1992).

Following the development of analytical methodologies to isolate and quantify SA, Malamy and co-workers (1990) observed that endogenous SA levels in TMV-resistant tobacco (Xanthi-nc) increased nearly 50-fold in TMV-inoculated leaves. Levels of SA increased 10-fold in uninoculated leaves of the TMV-inoculated plants. Following the increases in SA, PR-1 mRNA was observed to accumulate in proportion to salicylate levels. Feeding SA to excised Xanthi-nc tobacco leaves induced PR-1 proteins when endogenous SA levels were comparable to those observed during SAR (Yalpani *et al.*, 1991). In addition, increased levels of endogenous SA observed during SAR have recently been shown to correlate with enhanced resistance to TMV in hydroponically grown Xanthi-nc tobacco (Enyedi *et al.*, 1992), in an amphidiploid *Nicotiana debneyi* X *N. glutinosa* hybrid, as well as in older leaves of Xanthi-nc tobacco (Yalpani, Shulaev, and Raskin, unpublished). Increases in endogenous SA are also seen in tobacco after inoculation with necrotizing bacteria or fungi (Silverman and Raskin, unpublished).

The mobility of SA in plants following the resistance response has been investigated. Based on the physical properties of the molecule, a recently developed model for phloem mobility predicts that SA is nearly ideally suited for long distance phloem transport (Yalpani *et al.*, 1991). Analysis of the phloem sap of cucumber plants infected with *C. lagenarium* or tobacco necrosis virus showed that levels of SA increased dramatically between the time of infection and the establishment of SAR 7d later (Metraux *et al.*, 1990). Phloem exudates of TMV-inoculated Xanthi-nc tobacco leaves contained elevated levels of SA. The increase in exuded SA paralleled the observed increases in the excised leaf tissue (Yalpani *et al.*, 1991).

The distribution of SA in TMV-inoculated tobacco has also been investigated (Enyedi *et al.*, 1992). The levels of free SA were highest around the HR lesions. Hydrolytic treatments of leaf extracts revealed the presence of glucosyl-SA conjugate which accumulated primarily in the vicinity of the lesions (Enyedi *et al.*, 1992, Malamy *et al.*, 1992). The SA conjugate, tentatively identified as ß-O-D glucosylsalicylic acid, appeared to be non-mobile, and was not detectable in leaf exudates or uninfected leaves of TMV-inoculated plants. Treatment of tobacco with the SA-glucoside did not result in PR-protein induction nor in SAR (Enyedi, Yalpani, and Raskin, unpublished).

Currently, there are efforts underway in our laboratory to clone the genes which encode the enzymes responsible for SA biosynthesis and metabolism. Characterizing their expression and regulation will strengthen our understanding of plant disease resistance.

*   *   *   *   *

The determination of endogenous roles for SA has helped to explain its ubiquitous presence in plants. SA is a likely regulator of both disease resistance and thermogenesis. However, the pathway of signal transduction which leads from SA to these two different SA-regulated processes is not understood. Further research should elucidate other possible functions of SA, its mode of action, biosynthesis and metabolism.

REFERENCES

Antoniw, J.F., Ritter, C.E., Pierpoint, W.S., and van Loon, L.C., 1980, Comparison of three pathogenesis-related proteins from plants of two cultivars of tobacco infected with TMV, *J. Gen. Virol.* 47:79.

Elthon, T.E., and McIntosh, L., 1986, Characterization and solubilization of the alternative oxidase of *Sauromatum* mitochondria, *Plant Physiol.* 82:1.

Enyedi, A.J., Yalpani, N., Silverman, P., and Raskin, I., 1992, Localization, conjugation, and function of salicylic acid in tobacco during the hypersensitive reaction to tobacco mosaic virus, *Proc. Natl. Acad. Sci. USA* 89:2480.

Guedes, M.E.M., Richmond, S., and Kuc, J., 1980, Induced systemic resistance to anthracnose in cucumber as influenced by the location of the inducer inoculation with *Colletotricum lagenarium* and the onset of flowering and fruiting, *Physiol. Plant Pathol.* 17:229.

Herrmann, K., 1990, Salicylsaure und andere verbreitete Hydroxybenzoesauren und deren vorkommende Verbindungen in Lebensmitteln, *Ernahrungs-Umschau* 37:108.

Hiser, C., and McIntosh, L., 1990, Alternative oxidase of potato is an integral membrane protein synthesized *de novo* during aging of tuber slices, *Plant Physiol.* 93:312.

Horn, M.E., and Mertz, D., 1982, Cyanide-resistant respiration in suspension cultured cells of *Nicotiana glutinosa* L. *Plant Phsiol.* 69:1439.

Kapulnik, Y., Yalpani, N., and Raskin, I., 1992, Salicylic acid induces cyanide-resistant respiration in tobacco cell suspension cultures, *Plant Physiol.* (Submitted).

Kuc, J., 1982, Induced immunity to plant disease, *Bioscience* 32:854.

Linthorst, H.J.M., 1991, Pathogenesis-related proteins of plants, *Crit. Rev. Plant Sci.* 10:123.

Malamy, J., Carr, J.P., Klessig, D.F., and Raskin, I., 1990, Salicylic acid: a likely endogenous signal in the resistance response of tobacco to viral infection, *Science* 250:1002.

Malamy, J., Hennig, J., and Klessig, D.F., 1992, Temperature-dependent induction of salicylic acid and its conjugates during the resistance response to tobacco mosaic virus infection, *Plant Cell* 4:359.

Meeuse, B.J.D., and Raskin, I., 1988, Sexual reproduction in the arum lily family, with an emphasis on thermogenicity, *Sex. Plant Reprod.* 1:3.

Metraux, J.P., Signer, H., Ryals, J., Ward, E., Wyss-Benz, M., Gaudin, J., Raschdorf, K., Schmid, E., Blum, W., and Inverardi, B., 1990, Increase in salicylic acid at the onset of systemic acquired resistance in cucumber, *Science* 250:1004.

Raskin, I., 1992, Role of salicylic acid in plants, *Ann. Rev. Plant Physiol. Plant Mol. Biol.* 43:439.

Raskin, I., Ehmann, A., Melander, W.R., Meeuse, B.J.D., 1987, Salicylic acid - a natural inducer of heat production in Arum lilies, *Science* 237:1545.

Raskin, I., Skubatz, H., Tang, W., and Meeuse, B.J.D., 1990, Salicylic acid levels in thermogenic and non-thermogenic plants, *Ann. Bot.* 66:369.

Raskin, I., Turner, I.M., and Melander, W.R., 1989, Regulation of heat production in the inflorescences of an *Arum* lily by endogenous salicylic acid, *Proc. Natl. Acad. Sci. USA* 86:2214.

Rhoads, D.M., and McIntosh, L., 1991, Isolation and characterization of a cDNA clone encoding an alternative oxidase protein of *Sauromatum guttatum* (Schott), *Proc. Natl. Acad. Sci. USA* 88:2122.

Ross, A.F. 1961, Systemic acquired resistance induced by localized virus infections in plants, *Virology* 13:340.

Ross, A.F., 1966, Systemic effects on local lesion formation, *in*: "Viruses of Plants," A.B.R. Beemst, J. Bijkstra, eds., North Holland, Amsterdam.

Siedow, J.N., and Berthold, D.A., 1986, The alternative oxidase: a cyanide-resistant respiratory pathway in higher plants, *Physiol. Plant.* 66:569.

Van Herk, A.W.H., 1937, Die chemischen vorgange im *Sauromatum*-Koblen. II. Mitteilung, *Proc. Kon. Ned. Akad. Wet.* 40:607.

Ward, E.R., Uknes, S.J., Williams, S.C., Dincher, S.S., Wiederhold, D.L., Alexander, D.C., Ahl-Goy, P., Metraux, J.P., and Ryals, J.A., 1991, Coordinate gene activity in response to agents that induce systemic acquired resistance, *Plant Cell* 3:1085.

Weissmann, G., 1991, Aspirin, *Sci. Amer.* 264:84.

White, R.F., 1979, Acetylsalicylic acid (aspirin) induces resistance to tobacco mosaic virus in tobacco, *Virology* 99:410.

White, R.F., Antoniw, J.F., Carr, J.P., and Woods, R.D., 1983, The effects of aspirin and polyacrylic acid on the multiplication and spread of TMV in different cultivars of tobacco with and without the N-gene, *Phytopath. Z.* 107:224.

Yalpani, N., Silverman, P., Wilson, T.M.A., Kleier, D.A., and Raskin, I., 1991, Salicylic acid is a systemic signal and an inducer of pathogenesis-related proteins in virus-infected tobacco, *Plant Cell*, 3:809.

BLUE-LIGHT REGULATED GENE EXPRESSION

Lon S. Kaufman, Kathleen A. Marrs, Katherine M. F. Warpeha, Jie Gao, Keshab Bhattacharya, Judi Tilghman, and John F. Marsh III

Department of Biological Sciences
Laboratory for Molecular Biology
University of Illinois at Chicago
Chicago, Illinois 60680

INTRODUCTION

Plants rely heavily on light as an environmental cue. This is in addition to the use of light as an energy source via photosynthesis. The use of light to cue various developmental and/or environment adapting processes is referred to as photomorphogenesis. The occurrence of photomorphogenesis dictates that light is perceived and converted to a biochemical signal capable of eliciting the proper response. The components of the signal transduction mechanisms responsible for photomorphogenesis are poorly understood. We are interested in discerning the mechanism of blue-light regulated changes in gene expression.

The best studied and most obvious blue light responses are suppression of stem elongation and phototropic curvature (Baskin, 1986; Laskowski and Briggs, 1988; Warpeha and Kaufman, 1988, 1990). Both occur in response to a pulse of blue light and exhibit reciprocity. The two responses have different thresholds to blue light, suggesting two independent signal transduction mechanisms. Both responses occur in the stem and it is likely that the receptors and signal transduction mechanisms are housed within the stem.

Elongation and curvature are very complicated responses and it is difficult to determine where signal transduction stops and response initiates. Initiation of transcription is a definitive event that can be measured and is certain to be necessary for many photomorphogenic events. Thus, a study of the signal transduction mechanism responsible for blue-light-mediated transcription might prove useful in the identification of signal transduction components.

My lab has identified several nuclear gene families whose rate of transcription is effected by blue light. Blue light may act to increase the rate of transcription as is the case for the Cab gene family in pea and Arabidopsis, or may act to decrease the rate of transcription, as is the case for the pEA207 gene family in pea (Marrs and Kaufman, 1989, 1991; Gao and Kaufman unpublished). The effects of blue light on transcription are fluence dependent, occur immediately upon irradiation and occur in the absence of cytoplasmic protein synthesis. This indicates that the signal transduction mechanism is fully present in the cell prior to the blue light irradiation; the effects on Cab and pEA207 transcription are not the result of a gene cascade but represent the first effects on transcription.

A careful comparison of the rate of transcription with the steady state level of RNA reveals certain growth condition under which blue-light irradiation can induce RNA degradation. The effect on RNA turnover is exemplified by Cab RNA in red-light-grown pea seedlings irradiated with a single pulse of high-fluence blue light. As demonstrated in Fig. 1, the rate of Cab gene transcription increases immediately upon blue light irradiation. The rate of transcription reaches a plateau within three

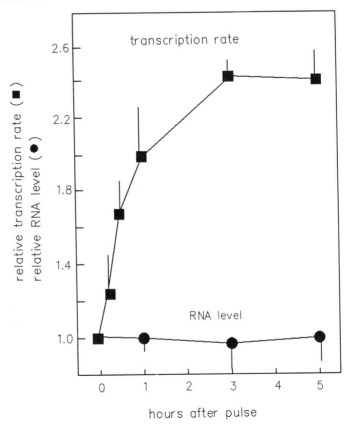

Figure 1. A comparison of the time course for Cab gene transcription and Cab RNA levels in six-day-old red-light-grown pea seedlings treated with a pulse of high fluence blue light.

hours and remains at the increased rate for at least 2 hours. Also shown in Fig. 1 is the steady state level of Cab RNA in comparably treated seedlings (Warpeha et al., 1989). Cab RNA does not accumulate under these conditions even though transcription is clearly occurring. These data strongly suggest that Cab RNA is degraded as a function of the blue-light treatment.

SIGNAL TRANSDUCTION MECHANISM

The mechanisms by which the blue-light signal is perceived, converted to a biological signal and transduced to the response machinery is poorly understood. Several attempts have been made to define the location and number of signal transduction mechanisms governing different blue light responses, as well as the components of these signal transduction mechanisms.

There are at least two locations containing one or several blue-light-initiated signal transduction mechanisms. For example, in pea, the mechanism regulating epicotyl elongation is located in the growing region of the stem (Gallagher et al., 1988; Short and Briggs, 1990) whereas the mechanism regulating Cab gene transcription is located in the apical bud (Warpeha et al., 1991; 1992).

The same tissue may contain more than one signal transduction systems. For example the rate of transcription for specific nuclear gene families in pea is regulated by two different blue light systems. The systems can be differentiated by the threshold fluence. The low-fluence system has a threshold at or below 10^{-1} umol m^{-2} while the high-fluence system has a threshold between 10^2 and 10^3 umol m^2 of blue light.

Different developmental stages of the same tissue may have different signal-transduction mechanisms. For example, in the case of transcription in the apical bud of pea, the low-fluence system is active in the apical bud of dark-grown and red-light-grown seedlings. However, the high-fluence system is only apparent in the developing leaves of the apical bud (Warpeha and Kaufman, 1991).

Recent work has concentrated on identifying components of these various blue-light-stimulated signal transduction mechanisms. This effort is ongoing in several labs and good progress has been made. However, as the various labs are examining the signal transduction mechanisms effecting different responses, often in different organisms, it is difficult to extrapolate interactions between the various components identified.

The Briggs group has identified a 117 kD polypeptide associated with plasma membranes derived from the growing region of pea epicotyls. This protein becomes phosphorylated upon irradiation of the purified membranes (Gallagher et al., 1988; Short and Briggs, 1990). The fluence-response data for the phosphorylation closely follows those of blue-light mediated suppression of epicotyl elongation in pea. It is possible that this protein is part of a complex containing the receptor or is the receptor itself. The 117 kD polypeptide does not appear to be in the apical bud or if present is not phosphorylated as a result of blue-light irradiation. This supports the notion of independent mechanisms for blue-light responses in the apical bud from those in the stem of pea.

Our group has identified a blue-light activated heterotrimeric GTP-binding regulatory protein associated with the plasma membrane of the apical buds of etiolated pea (Warpeha et al., 1991). Irradiation of the purified membranes results in activation of the G-protein.

The alpha subunit of the G-protein appears to be a 40 kD polypeptide. This polypeptide is identified by antibody generated against transducin, is ADP-ribosylated by pertussis toxin in the absence of blue light but not after irradiation, is ADP ribosylated by cholera toxin only in the presence of blue-light and GTP, and will bind a UV-crosslinking, radiolabeled, GTP analog only upon irradiation with blue light. We are currently in the process of isolating this protein.

Sarah Assmann's group, using pertussis and cholera toxin treatments, has identified a possible role for a G-protein in the blue-light mediated process of stomatal opening (Fairley-Grenot and Assmann, 1991).

The receptors driving the various blue-light-responsive signal transduction mechanisms have not been identified, indeed controversy still persists over the nature of the chromophore. In collaboration with the Briggs group we have identified a flavin as the chromophore of the receptor responsible for the activation of the G-protein. Heterotrimeric GTP-binding regulatory proteins couple directly to a transmembrane receptor. Therefore, any effect of inhibitors of flavin action on G-protein activity would indicate that the receptor is a flavoprotein. The results of such experiments indicate that the blue-light receptor requires excitation of and energy transfer from a flavin (Warpeha et al., 1992). Furthermore, the experiments suggest that the flavin functions through a triplet state.

A similar set of experiments using membranes derived from the pea stem and the phosphorylation event as the response indicate the involvement of a flavin (Short et al., 1992). Although unlikely, it remains possible that the flavin is not within the receptor as the proximity of the phophorylated protein to the receptor is not officially known. The flavin probably functions via a singlet state. As suggested earlier the signal transduction mechanism regulating stem elongation is probably physically distinct from that effecting transcription in the apical bud. These newer data suggest that the receptors are biochemically distinct.

The mechanism by which blue light modulates transcription is unknown. We have identified several nuclear gene families as being blue-light regulated (Marrs and Kaufman, 1989, 1991) and in the case of the Cab gene family in pea and Arabidopsis we have identified specific members which are regulated (Marrs, Gao and Kaufman unpublished data). Nuclear extracts derived from blue-light treated seedlings exhibit altered (slowed) mobility shifts as compared with extracts derived from untreated seedlings when the DNA probes are derived from blue-light regulated members of the Cab gene family. Altered shifts are not apparent when probes derive from members of the Cab gene family not regulated by blue light (Marrs 1991; Marrs, Tilghman and Kaufman, unpublished observation).

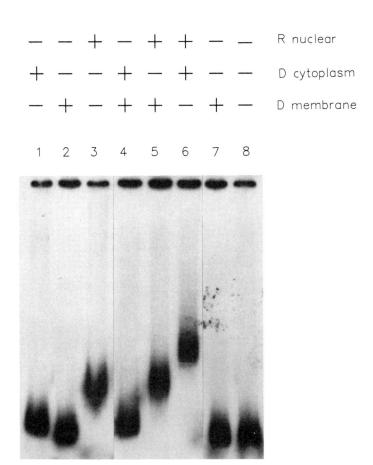

Figure 2. Mobility shift assays using a 470 bp Cab DNA probe from pea and various pea cell fractions, alone and in combination. Refer to text for full explanation.

The blue-light receptor is likely to be located in the plasma membrane. As a consequence, the "signal" must travel into the nucleus. The protein responsible for carrying the "signal" into the nucleus may 1) be active prior to the blue-light treatment and enter the nucleus as a function of blue light, 2) have free access to the nucleus but require blue light for activation, 3) require blue light for both activation and access to the nucleus.

The first possibility can be discerned from the latter two by mixing cytoplasm from unirradiated seedlings with nuclear extract from unirradiated seedlings. If the cytoplasm has an effect on the mobility shift then it is possible that the signal carrier is present in the active state in the cytoplasm and that the blue light acts to allow the protein access to the nucleus.

To test this hypothesis we have used a 470 bp fragment derived from a blue-light regulated member of the pea Cab gene family in mobility shift assays (see Fig. 2). The probe contains approximately 440 bp of upstream sequence. When mixed with plasma membranes from etiolated seedlings (lanes 2 and 7), cytoplasm from etiolated seedlings (lane 1), or a combination the two (lane 4), there is no apparent mobility shift (see lane 8 for free probe). Nuclear extract derived from nuclei purified from red grown seedlings exhibits a mobility shift (lane 3) which is unaffected by the addition of the plasma membranes (lane 5). However, when cytoplasm is added to this mix there is a significant change in the mobility of the fragment (lane 6). The data shown is derived from shift assays performed in the absence of blue light. Thus light is not necessary to achieve the altered shift observed in lane 6.

These data indicate that a protein exists within the cytoplasm derived from the apical buds of etiolated seedlings that is capable of altering the shift of the 470 bp fragment. The protein itself is not capable of altering the mobility of the DNA fragment as use of the cytoplasmic extract alone does not result in a shift. Thus, the activity of this protein is dependent upon the presence of the red nuclear fraction.

It is possible that the role of blue light in the transcription process is to allow access to the nucleus for this and/or other proteins present in the cytoplasm. For example, blue light irradiation might result in the opening of a pore in the nuclear envelope. This conjecture is shown schematically in Fig. 3.

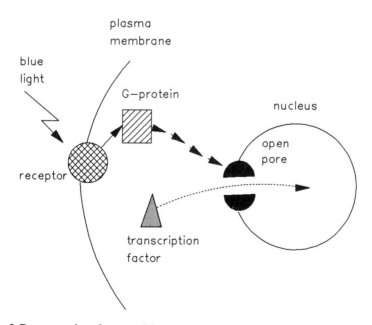

Figure 3. Representation of one possible mechanism for blue light regulation of transcription.

REFERENCES

Baskin, T.I., 1986, Redistribution of growth during phototropism and nutatin in the pea epicotyl, Planta 169:406.

Gallagher, S., Short, T.W., Ray, P.M., Pratt, L.H., and Briggs, W.R., 1988, Light-mediated changes in two proteins found associated with plasma membrane fractions from pea stem sections Proc. Natl. Acad. Sci. USA 85:8003.

Fairley-Grenot, K., and Assmann, S.M., 1991, Evidence for G-protein regulation of inward K+ channel current in guard cells of Fava bean, Plant Cell 3:1037.

Laskowski, M., and Briggs, W.R., 1988, Regulation of pea epicotyl elongation by blue light: Fluence response relationships and growth distribution, Plant Physiology 89:293.

Marrs, K.A., and Kaufman, L.S., 1989, Blue light regulation of transcription for nuclear-coded genes in pea, Proc. Natl. Acad. Sci. USA 86:4492.

Marrs, K.A., 1991, Transcriptional regulation of nuclear genes in Pisum sativum by blue light, PhD. Thesis.

Marrs, K.A., and Kaufman, L.S., 1991, Rapid transcriptional regulation of Cab and pEA207 gene families by blue light in the absence of protein synthesis, Planta 183:327.

Short, T.W., and Briggs, W.R., 1990, Characterization of a rapid, blue light mediated change in detectable phosphorylation of a plasma membrane protein from pea (Pisum sativum) seedlings, Plant Physiology 92:179.

Short, T.W., Porst, M., and Briggs, W. R., 1992, A photoreceptor system regulating in vivo and in vitro phosphorylation of a pea plasma membrane protein, Photochem. Photobiol. In press.

Warpeha, K.M.F., and Kaufman, L.S., 1988, Blue light regulation of epicotyl elongation in pea, Plant Physiology 89:544.

Warpeha, K.M.F., and Kaufman, L.S., 1990, Two distinct blue light responses regulate epicotyl elongation in pea, Plant Physiology 92:495.

Warpeha, K.M.F., Hamm, H.E., Rasenick, M.M., and Kaufman, L.S., 1991, A blue-light activated GTP binding protein in the plasma membranes of etiolated peas, Proc. Natl. Acad. Sci. USA 88:8925.

Warpeha, K.M.F., Marrs, K.A., and Kaufman, L.S., 1989, Blue light regulation of specific nuclear coded transcripts in pea, Plant Physiology 91:1030.

Warpeha, K.M.F., Kaufman, L.S., and Briggs, W.R., 1992, A flavoprotein may mediate the blue light activated binding of guanosine 5'-triphosphate to isolated plasma membranes of Pisum sativum, Photochem. Photobiol. 55:595.

THE ROLE OF THE MAIZE *VIVIPAROUS-1* GENE IN REGULATION OF SEED MATURATION

Donald R. McCarty

Horticultural Sciences Department
University of Florida
Gainesville, FL 32611

INTRODUCTION

Late in the course of seed formation tissues that are destined to remain viable in the dry seed undergo a maturation process during which further development is arrested and tolerance to desiccation is acquired. The viviparous mutants of maize which fail to complete maturation identify genes that are essential for this process (Robertson, 1955; McCarty and Carson, 1991; McCarty et al., 1992). In this paper we will address the regulation of the maturation program on two levels: 1) how intrinsic and extrinsic signals are integrated to produce a developmentally specific response and 2) how the diverse metabolic pathways associated with maturation are integrated by a regulatory hierarchy into a common developmental program. We explore the notion that these two integration processes are intimately related and involve a common mechanism operating at the level of transcriptional regulation. This model is motivated by our analysis of the role of the *Viviparous-1 (Vp1)* gene in regulating the maturation program in maize. Finally, our results offer some insight into how at least part of the regulatory network controlling maturation evolved.

INTEGRATION OF EXTRINSIC AND EXTRINSIC DEVELOPMENTAL SIGNALS

Maturation is not an intrinsic feature of plant embryogenesis (Quatrano, 1987). Embryos that are removed from the developing seed at an early stage and placed in culture complete the morphogenetic program and develop into plants without entering a period of developmental arrest or acquiring desiccation tolerance. Embryos induced to form from somatic cells in tissue culture also develop into plantlets without going through a distinct maturation phase. It is, in fact, difficult to establish an arrested, desiccation tolerant state in plant embryos grown *in vitro* (Senaratna et al., 1989). That embryo development outside of the seed is viviparous implies that extrinsic factors present in the environment of the developing seed are essential for maturation. The viviparous mutants of maize which cause or allow viviparous development to occur within the context of the seed are one means of identifying such factors.

Abscisic acid (ABA) is strongly implicated as a key extrinsic regulator of maturation by the genetics of vivipary in maize (reviewed in McCarty and Carson, 1990; McCarty, 1992), as well as a growing body of molecular and biochemical evidence (reviewed in Skriver and Mundy, 1990). Most of the viviparous mutants of maize act by blocking synthesis of ABA (Neil et al., 1987; Robichaud et al., 1980). The exceptional mutant is *vp1* which has normal ABA metabolism (Robichaud and Sussex, 1987), but exhibits a seed specific insensitivity to the hormone (Robichaud et al., 1980). A number of maturation associated genes have been isolated and shown to be regulated by ABA during embryogenesis. Among the most significant of these are genes belonging to the LEA class and the 7S globulin storage proteins (Skriver and Mundy, 1990). Promoter sequences involved in ABA regulation of Em, a LEA gene isolated from wheat, have been identified Marcotte et al. (1988; 1989) and characterized (Guiltinan et al., 1991).

In addition to being expressed during normal embryogenesis, many of ABA regulated genes associated with maturation can be activated by ABA or drought stress in non-seed tissues of the plant (Gomez et al., 1988; Mundy et al., 1988; Marcotte et al., 1988). However, while components of the program are expressed or induced in various plant tissues at other stages in the plant life cycle by external hormone or osmotic treatments, the complete process leading to developmental arrest and desiccation tolerance occurs exclusively in the developing seed. Neither the intrinsic or extrinsic developmental signals by themselves are sufficient to specify the complete maturation program. ABA must in some way be integrated with other intrinsic developmental information to induce maturation. The *vp1* mutant which specifically affects the ABA response in the seed may identify a gene involved in this integration process.

The *vp1* gene has been cloned by transposon tagging (McCarty et al., 1989). We have recently shown the VP1 protein has properties of a transcriptional factor (McCarty et al., 1991). Over-expression of VP1 in maize protoplasts is sufficient to strongly activate transcription from the promoter of the wheat Em gene. In combination, VP1 and exogenous ABA interact synergistically in regulating the Em gene (McCarty et al., 1991). In addition to blocking the arrest of embryo development and activation of Em, the *vp1* mutant prevents activation of the anthocyanin biosynthetic pathway in seed tissues during maturation (Robertson, 1955). Recent analysis of *Vp1*'s role in regulating the latter pathway has revealed part of a regulatory network controlling maturation.

THE REGULATORY HIERARCHY INTEGRATING THE ANTHOCYANIN PATHWAY INTO MATURATION

While maturation entails complex changes in gene expression, many of the genes and pathways involved are not strictly specific to maturation or even the ABA response. A striking illustration of this point is the accumulation of purple anthocyanin pigments in the aleurone layer of the endosperm and scutellum tissue of the embryo during seed maturation in maize. The fact that mutations in the *vp1* gene prevent both developmental arrest and activation of the anthocyanin pathway indicates that both pathways are regulated by a common mechanism (McCarty and Carson, 1990). Because genetic control of the anthocyanin pathway has been studied extensively in maize, this pathway is a particularily useful model for dissecting the regulatory mechanisms responsible for integrating a general purpose pathway into a specific developmental program. It is known, for example, that the anthocyanin pathway includes a common set of structural genes that are active in diverse plant tissues at many different times in the plant life cycle. In recent years, several of the regulatory genes responsible for developmental control of the pathway have been identified by mutation and cloned (reviewed in Dooner et al., 1991). In addition to *Vp1* two other regulatory genes, *C1* and *R1*, are required for activation of the anthocyanin pathway in the seed. Unlike *vp1*, the *c1* and *r1* mutants affect only anthocyanin pigmentation. The *C1* and *R1* genes encode transcription factors with sequence homology to the *myb* and *myc* proteins in animals, respectively. Functional studies indicate that the C1 and R1 proteins interact to activate transcription of the structural genes in the pathway (Goff et al., 1991). What concerns us here is how this regulatory network specifically associated with the anthocyanin pathway is subsumed by or integrated into the regulatory hierarchy controlling seed maturation.

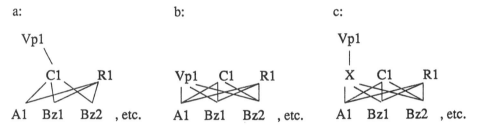

Figure 1. Three alternative models for the interaction of the *Vp1*, *C1* and *R1* regulatory genes for the anthocyanin pathway in maize. *A1*, *Bz1* and *Bz2* are structural genes in the pathway. A. *Vp1* is a higher level regulator of the *C1* gene (see text). *C1* and *R1* interact directly with promoters of structural genes. B. All three regulatory genes act in parallel and interact directly with the structural genes. C. *Vp1* is required for expression of a fourth unidentified factor.

The phenotypes of the regulatory mutants suggest that because it is more pleiotropic, *Vp1* is likely to act earliest in the pathway, perhaps occupying a higher level position in a hierarchy controlling of expression of *C1* and/or *R1* in the seed (Fig. 1a). However, other models are possible. For example, VP1 might interact directly with the structural genes in the anthocyanin pathway (Fig. 1b) or activate yet another unidentified regulatory gene (Fig. 1c). These models can, however, be distinguished using a reverse genetics approach. A key prediction of the hierarchy model is that constitutive overexpression of the two downstream regulatory genes, *C1* and *R1*, should be sufficient to activate the pathway and complement the anthocyanin deficiency in the *vp1* mutant. The alternative models depicted in Fig.1 b&c do not predict complementation in this situation. To test this prediction we used particle-gun mediated transformation to introduce gene constructs containing the respective C1 and B1 coding sequences (B1 is one member of the *R1* gene family) fused to the constitutive cauliflower mosaic virus 35S promoter into the exposed aleurone of *vp1* mutant kernels (Hattori et al., 1992). Goff et al. (1990, 1991) had previously shown that the 35S-B1 and 35S-C1 gene constructs could complement anthocyanin synthesis when introduced into *C1 r1* and *c1 R1* tissues, respectively. Transformation of *vp1 C1 R1* aleurone cells with both the 35S-C1 and 35S-B1 plasmids induced anthocyanin pigment synthesis within 12h following bombardment firmly supporting the hierarchy model in Fig. 1a (Hattori et al., 1992). A direct interaction of VP1 with anthocyanin structure genes is not essential.

Is *Vp1* required for expression of just one or both of the anthocyanin regulatory genes? To address this question we asked whether introduction of either the 35S-C1 or 35S-B1 plasmids individually was sufficient to complement the *vp1* deficiency (Hattori et al., 1992). The clear result was that expression of C1 alone was sufficient to activate the pathway, whereas, the B1 construct by itself was ineffective. This result indicated that it is specifically the C1 function which limits the anthocyanin pathway in the *vp1* mutant. The *R1* gene is apparently already active in this tissue. The implication is that *Vp1* controls the anthocyanin pathway primarily by regulating *C1*. In accord with this model, we find that transcripts of the *C1* gene are absent in *vp1* mutant tissues (McCarty et al., 1989).

Having established a regulatory link between *Vp1* and *C1*, it is informative to ask whether the anthocyanin pathway can be genetically uncoupled from maturation. The block in anthocyanin expression in the *vp1* mutant is actually conditional and can be overridden by prolonged exposure of the viviparous seed to light (McCarty and Carson, 1990). Chen and Coe (1978) described an allele of *c1*, designated *c1-p*, which produces an analogous effect. Like seed that carry a standard null mutant *c1* allele (e.g. *c1-n*) *c1-p* seed are colorless at maturity; however, if exposed to light during their development *c1-p* seed become pigmented during germination. Chen and Coe (1978) further established that this light dependent expression of *c1-p* during germination did not require *Vp1* by constructing the *c1-p, vp1* double mutant. A key observation is that wild-type *C1* alleles and *c1-p* exhibit essentially similar behavior when placed in a *vp1* mutant background (McCarty and Carson, 1991). What distinguishes *C1* and *c1-p* alleles then is their ability to respond to *Vp1* during seed maturation. The *c1-p* allele apparently encodes a functional product, but its regulation is altered such that it is uncoupled from *Vp1* and maturation. Comparison of the *C1* and *c1-p* promoter DNA sequences in the 400 bp region immediately upstream of the transcription start site revealed a single 5 bp sequence polymorphism which distinguishes all known wild-type promoters from *c1-p*. *C1* type promoters contain a direct repeat of the sequence GTGTC beginning at -153 which is present in a single copy in *c1-p*.

CIS-ACTING SEQUENCES INVOLVED IN VP1 AND ABA REGULATION OF TRANSCRIPTION.

The cis-acting sequences responsible for regulation of the *C1* gene were further resolved using a transient expression system based on electroporation of maize protoplasts (Hattori et al., 1992). A reporter gene was constructed containing the 5′ flanking DNA sequences from a wild-type *C1* gene fused to the bacterial GUS gene. (An intron sequence from the *Sh1* gene was also included to enhance expression of this gene in the protoplast system (Vasil et al., 1989)). We discovered that this *C1*-Sh-GUS gene could be activated (10 to 14 fold) by incubation of the protoplasts with micromolar concentrations of ABA following electroporation. Moreover, overexpression of the VP1 protein (induced by simultaneously transforming the protoplasts with a CaMV 35S-Sh1-Vp1 construct) caused a 5 to 7 fold activation of *C1*-Sh-GUS. In an interesting contrast, however, to what we had previously observed with the wheat Em gene (McCarty et al., 1991), ABA and VP1 do not interact synergistically at *C1*. Used in combination VP1 over-expression and ABA produce no greater activation than ABA alone. By generating and testing a comprehensive series of deletions in the *C1* promoter we were able to delineate a 27 bp sequence between -157 and -130 that is essential for both ABA regulation and trans-activation by VP1 in the protoplast assay. It is significant that this sequence includes the GTGTC repeat implicated by the *c1-p* polymorphism. This fact argues strongly that the response obtained in protoplasts is relevant to what is happening in the developing seed.

When the *c1-p* promoter sequence is tested in the protoplast system, we find that it is not activated by ABA, but can still be weakly trans-activated by VP1. Mutations in the adjacent downstream portion of the 27 bp block abolish both responses. Thus, the 27 bp region can be resolved into a sequence more or less specific for ABA regulation and a sequence required for both ABA and VP1 regulation.

One can ask whether there is any discernible relationship between the cis-regulatory elements identified by these experiments in the *C1* promoter and sequences that have been implicated in ABA regulation of other genes. In particular, Marcotte et al. (1989) have identified an element containing the sequence CACGTGGC (designated Em1a) that is involved in ABA regulation of the wheat Em gene (Guiltinan et al., 1990). This sequence bears little obvious similarity to sequences in the 27 bp region of the *C1* promoter. Inspection of the *C1* promoter, however, turned up two motifs that do closely resemble the Em1a element. Deletion of either one or both copies of this sequence had little effect on ABA activation of *C1*. This raises the possibility that though Em and *C1* are both regulated by ABA, different transcription factors may be involved. The situation is further complicated by the fact that the Em1a sequence, itself, is a member of the widely distributed "G-box" family of regulatory elements that have been implicated in light and anaerobic regulation of gene expression (Guiltinan et al., 1990; Armstrong et al., 1992; DeLisle and Ferl, 1990). Presence or absence of the Em1a sequence does not, by itself, correlate well with regulation by ABA.

On the otherhand the region in the *C1* promoter that is required for both ABA and VP1 activation includes a sequence motif (TCCATGCATGCAC) which also appears to be conserved in the 5′ flanking DNA of several ABA regulated genes isolated from grasses (Hattori et al., 1992). This sequence is closely related to the CATGCATG motif that appears in a variety of seed specific plant genes (Dickinson et al., 1988). While the regulatory role of this motif in gene expression is still uncertain, it seems unlikely that it is specific to ABA or VP1 regulation.

To summarize, there would not appear to be a single cis-acting element that is specific to ABA or VP1 regulated gene expression. Em and *C1* appear to require different combinations of regulatory sequences and may activated by different mechanisms. This difference in mechanism may contribute to the strikingly different interaction seen between VP1 trans-activation and ABA regulation of these two genes. VP1 and ABA are strongly synergistic in activating Em (McCarty et al., 1991) whereas, the interaction at *C1* is less than additive (Hattori et al., 1992). A single transcriptional activator, VP1, nonetheless is evidently able to recognize both motifs. These observations suggest a model for VP1 action that may offer some insight into the mechanism by which signals and pathways are integrated in plant development.

INTEGRATION OF PATHWAYS AND SIGNALS IN MATURATION

The observation that regulatory sequences in *C1* can be resolved into an ABA specific region and a region required for both ABA and VP1 activation suggests that multiple transcription factors may interact to cause activation. VP1 contains a potent transcriptional activation sequence (McCarty et al., 1991) and is therefore likely to be an essential component of such a complex. However, because recombinant VP1 lacks a detectable sequence specific DNA binding activity (McCarty et al., 1991; Hattori et al., 1992), we speculate that its participation in this complex may be strongly dependent on protein-protein interactions. In this scenario, binding of VP1 with its potent transcriptional activator to the *C1* promoter would be dependent on the presence of at least two other DNA binding proteins. While the activity of the upstream factor may be modulated in some way by ABA, the second factor need not be specifically associated with ABA signal transduction, but could be tied to a different developmental signal. (Recall that the CATGCATG motif has been implicated in seed specific expression (Dickinson et al., 1988)). To the extent that recruitment of VP1 to the *C1* promoter requires two or more signals in combination formation of this complex could be viewed as an integration process. In this light, the seed-specific, hormone-insensitive phenotype of the *vp1* mutant is viewed as a direct consequence of the failure to integrate the ABA signal transduction pathway with other intrinsic developmental signals.

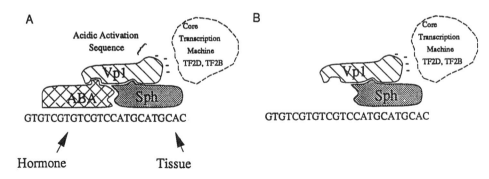

Figure 2. A model for integration of developmental signals controlling *C1* transcription. A. VP1 interacts with two other DNA binding proteins: one that is associated with ABA signal transduction and a second possibly specifying intrinsic developmental information. The acidic activation sequence (McCarty et al., 1991) in VP1 interfaces with the core transcriptional machinery. B. If present in excess VP1 can activate *C1* (or *c1-p*) in the absence of ABA by interacting with the Sph factor alone.

In the above model we envision VP1 as an activator protein able to specifically interact with several other proteins. It is this multivalent character in combination with a strong transcriptional activation function that allows VP1 to serve an integrating function. We suggest that VP1 activation of other downstream genes, such as Em, involves interactions with a somewhat different set of transcription factors, perhaps bound to different cis-acting sequences. An interesting consequence of this model is that the polyvalent structure that allows integration of developmental signals also provides a mechanism for bringing diverse pathways under control of a single regulatory molecule.

A model is useful only to the extent that it makes testable predictions. If the various protein-protein interactions involve discrete regions in the VP1 protein it should be possibe to make structural mutations in VP1 that differentially affect its ability to activate Em and *C1*. In fact, at least one class of *vp1* mutations appear to have just this property. Four independent *vp1* mutations are known which block anthocyanin synthesis, but produce nearly normal desiccation tolerant seed (McCarty et al., 1989a). These mutants are apparently able to activate a substantial component of the maturation pathway, but not *C1* and the anthocyanin pathway. At least one of these mutants, *vp1-McWhirter*, encodes a structurally altered protein (McCarty et al., 1989a; McCarty et al., 1991; C. B. Carson and D. R. McCarty, unpublished data). This phenotype contrasts with other alleles, such as *vp1-mum3* and *vp1-mum5* which appear to be simply leaky. The latter mutants produce reduced levels of VP1 and quantitatively reduce both pigment expression and the supression of vivipary (McCarty, 1992; C. B. Carson and McCarty, unpublished data). The availability of a rapid transient expression assay for VP1 activity using different downstream promoters as reporters will allow further resolution of functional regions within the VP1 protein.

EVOLUTION OF A REGULATORY NETWORK

In addition to serving as a model for how pathways can be integrated into a common developmental program, the anthocyanin pathway offers some insight into how the regulatory networks that achieve this integration may have evolved. The origin of *c1-p*, the allele which causes the anthocyanin regulation to be uncoupled from maturation is relevant to that question. Because *c1-p* occurs as a natural variant in maize populations (Chen and Coe, 1978) we cannot be certain whether presence (*C1*-like) or absence *(c1-p*-like) of the GTGTC direct repeat in the *C1* promoter sequence is the ancestral form. It is tempting, however, to speculate that the *c1-p* form is actually the progenitor of the *C1* promoter sequence. Synthesis of anthocyanins in the aleurone during seed development is a characteristic of maize not found in *Teosinte* species that are probably ancestral to maize (John Doebley, personal communication). This would suggest that the integration of the anthocyanin pathway into maturation may have occured relatively recently in the evolution of maize. Polymorphisms consisting short direct duplications frequently distinguish alleles of maize genes (Zach et al., 1987). Sequence alterations of this kind are typical of sites that have been visited by transposable elements which are evidently very active in the maize genome.

Regardless of the direction of the mutation, this is an elegant example of how a simple alteration in a regulatory gene has produced a new pattern of developmental regulation for an entire pathway. It is parsimonous for the evolution to have occured in the cis-acting sequence controlling expression of a single trans-acting factor rather than in the promoter of each of the more numerous structural genes. Because even subtle changes in expression of a regulatory gene are potentially amplified by the cascade effect to produce large changes in the expression of a pathway, mutations in the promoter of a regulatory gene may also be more likely to produce an altered phenotype. It is interesting, in this respect, that *C1* is much less strongly activated by VP1 and ABA (ca. 10-fold) in protoplasts than is Em (ca. 1000-fold) (McCarty et al., 1991). Nevertheless, if we make reasonable assumptions for the kinetics of activation of a structural gene by the C1 protein, it can be shown that a 10-fold change in *C1* expression is sufficient produce a 50-fold activation of the structural gene (see Hattori et al., 1992). This sensitivity exists, because

C1 is a very potent activator of the structural genes in the pathway (Goff et al., 1991). Ironically, the strong coupling between the structural genes and their immediate regulator may facillitate the evolution of new developmental patterns by amplifying weak couplings between regulatory genes.

ACKNOWLEDGEMENTS

This work was supported by grants from the National Science Foundation and the McKnight foundation.

REFERENCES

Armstrong, G. A., Weishaar, B. and Hahlbrock, K. 1992. Hommodimeric and heterodimeric leucine zipper proteins and nuclear factors from parsley recognize diverse promoter elements with ACGT cores. Plant Cell. 4:525-537.

Chen, S. M. and E. H. Coe, Jr. 1978. Control of anthocyanin synthesis by the *C* locus in maize. Biochem. Genet. 15: 333-346.

Coe, E. H. and M. G. Neuffer. 1977. The genetics of corn. In Corn and corn improvement. (G. F. Sprague, ed). pp. 111-213. American Society of Agronomy, Madison, WI.

Cone, K. C., F. A. Burr and B. Burr. 1986. Molecular analysis of the maize anthocyanin regulatory locus *c1*. Proc. Nat. Acad. Sci. USA. 83: 9631-9635.

DeLisle, A. and Ferl, R. J. 1990. Characterization of the *Arabidopsis Adh* G-box binding factor. Plant Cell. 2: 547-557.

Dooner, H. K. 1985. *Viviparous-1* mutation in maize conditions pleiotropic enzyme deficiencies in the aleurone. Plant Physiol. 77:486-488.

Dickinson, C. D., R. P. Evans and Niels C. Neilsen. 1988. RY repeats are conserved in the 5'-flanking regions of legume seed-protein genes. Nucleic Acids Res. 16: 371.

Goff, S., K. C. Cone and M. E. Fromm. 1991. Identification of functional domains in the maize transcriptional activator C1: comparison of wild-type and dominant inhibitor proteins. Genes and Dev. 5:298-309.

Goff, S., T. M. Klein, B. A. Roth, M. E. Fromm, K. C. Cone, J. P. Radicella and V. L. Chandler. 1990. Transactivation of anthocyanin biosynthetic genes following transfer of *B* regulatory genes into maize tissues. EMBO J. 9:2517-2522.

Gomez, J., Sanchez-Martinez, D., Stiefel, V., Rigau, J., Puigdomenech and Pages, M. 1988. A gene induced by the plant hormone abscisic acid in response to water stress encodes a glycine-rich protein. Nature. 334:262-264.

Guiltinan, M. J., W. R. Marcotte, R. S. Quatrano. 1990. A leucine zipper protein that recognizes an abscisic acid response element. Science 250:267-270.

Hattori, T., Vasil, V., Rosenkrans, L., Hannah, L. C., McCarty, D. R. and Vasil, I. K. 1992. The *Viviparous-1* gene and abscisic acid activate the *C1* regulatory gene for anthocyanin biosynthesis during seed maturation in maize. Gen. Dev. 6:609-618.

Ludwig, S. R., L. F. Habera, S. L. Delaporta and S. R. Wessler. 1989. *Lc*, a member of the maize *R* gene family responsible for tissue-specific anthocyanin production, encodes protein similar to transcriptional activators and contains the *myc*-homology region. Proc. Natl. Acad. Sci. USA 86:7092-7096.

Marcotte, W. R. Jr., C. C. Bayley and R. S. Quatrano. 1988. Regulation of a wheat promoter by abscisic acid in rice protoplasts. Nature 335: 454-457.

Marcotte, W. R. Jr., S. H. Russell and R. S. Quatrano. 1989. Abscisic acid response sequences from the Em gene of wheat. Plant Cell 1: 969-976.

McCarty, D. R. 1992. The role of VP1 in regulation of seed maturation in maize. Biochem. Soc. Trans. 20:89-92.

McCarty, D. R., C. B. Carson, M. Lazar and S. C. Simonds. 1989a. Transposable element induced mutations of the *viviparous-1* gene of maize. Dev. Genetics. 10: 473-481.

McCarty, D. R., C. B. Carson, P. S. Stinard and D. S. Robertson. 1989b. Molecular analysis of *viviparous-1*: An abscisic acid insensitive mutant of maize. Plant Cell 1: 523-532.

McCarty, D. R. and C. B. Carson. 1990. The molecular genetics of seed maturation in maize. Physiol Plant. 81: 267-272.

McCarty, D. R., T. Hattori, C. B. Carson, V. Vasil and I. K. Vasil. 1991. The *viviparous-1* developmental gene of maize encodes a novel transcriptional activator. Cell 66: 895-905.

Mundy, J., K. Yamaguchi-Shinozaki and N.-H. Chua. 1990. Nuclear proteins bind conserved elements in the abscisic acid responsive promoter of a rice *rab* gene. Proc. Nat. Acad. Sci. USA. 87:406-410.

Neill, S. J., R. Horgan and A. D. Parry. 1986. The carotenoid and abscisic acid content of viviparous kernels and seedlings of *Zea mays L.* Planta 169:87-96.

Paz-Ares, J., D. Ghosal, U. Wienand, P. Peterson and H. Saedler. 1987. The regulatory locus *c1* of *Zea mays* encodes a protein with homology to *myb* proto-oncogene products and with structural similarities to transcriptional activators. EMBO J. 6: 3553-3558.

Robertson, D. S. 1955. The genetics of vivipary in maize. Genetics 40: 745-760.

Robichaud, C. S., J. Wong and I. M. Sussex. 1980. Control of *in vitro* growth of viviparous embryo mutants of maize by abscisic acid. Dev. Genetics 1: 325-330.

Robichaud, C. S. and I. M. Sussex. 1986. The response of *viviparous-1* and wildtype embryos of *Zea mays* to culture in the presence of abscisic acid. J. Plant Physiol. 126: 235-242.

Robichaud, C. S. and I. M. Sussex. 1987. The uptake and metabolism of ^{14}C-ABA by excised wild type and viviparous-1 embryos of Zea mays L. J. Plant Physiol. 130: 181-188.

Roth, B. A., S. A. Goff, T. M. Klein and M. E. Fromm. 1991. *C1* and *R1* dependent expression of the maize *Bz1* gene requires sequences with homology to mammalian *myb* and *myc* binding sites. Plant Cell 3:317-325.

Senaratna, T., McKersie, B. D. and Bowley, S. R. 1989. Desiccation tolerance of alfalfa (Medicago sativa L.) somatic embryos. Influence of abscisic acid, stress pretreatments and drying rates. Plant Sci. 65:253-259.

Skriver, K. and J. Mundy. 1990. Gene expression in response to abscisic acid and osmotic stress. The Plant Cell 2: 503-512.

Vasil, V., M. Clancy, R. J. Ferl, I. K. Vasil and L. C. Hannah. 1989. Increased gene expression by the first intron of the maize *shl* locus in grass species. Plant Physiol. 91:1575-1579.

Zack, C. D., R. J. Ferl and L. C. Hannah. 1986. DNA sequence of a *shrunken* allele of maize evidence for visitation by insertional sequences. Maydica 31: 5-16.

LOVASTATIN INDUCES CYTOKININ DEPENDENCE IN TOBACCO CULTURES

Dring N. Crowell and Michael S. Salaz

Department of Biology
Indiana University-Purdue University at Indianapolis
723 W. Michigan Street
Indianapolis, IN 46202-5132

ABSTRACT

The mevalonate pathway participates in the formation of cytokinins, abscisic acid, gibberellins, ubiquinones, plastoquinones, dolichols, sterols, carotenoids, chlorophylls, and numerous other isoprenoids in plant cells. The first committed step in this pathway is the irreversible reduction of 3-hydroxy-3-methylglutaryl CoA (HMG CoA) to mevalonate, a reaction that is catalyzed by HMG CoA reductase. We have studied the effects of a potent HMG CoA reductase inhibitor (lovastatin) on growth of a cytokinin-independent *Nicotiana tabacum* cell line and found that growth inhibition at low concentrations of inhibitor is reversed by various cytokinins. However, cytokinin analogs with little or no cytokinin activity do not restore growth to lovastatin-treated cells. These results are consistent with the hypothesis that cytokinin biosynthesis is more sensitive to lovastatin than the biosynthesis of other essential isoprenoid compounds in tobacco cells.

INTRODUCTION

Cytokinins have profound effects on plant growth and development (15, 16). For example, cytokinins stimulate chloroplast and lateral bud development (12) and suppress tissue senescence (18, 19). In addition, cytokinins promote cell division (15, 16) and are involved in tumorous diseases of plants (1).

Cytokinin is produced by the transfer of an *iso*-pentenyl group from dimethylallyl pyrophosphate to AMP and is, therefore, a product of the mevalonate pathway (see Figure 1) (1, 3, 5, 8, 9, 13). HMG CoA reductase, which catalyzes mevalonate biosynthesis, is tightly regulated in both plant and mammalian systems (phytochrome, phytohormones and other factors have

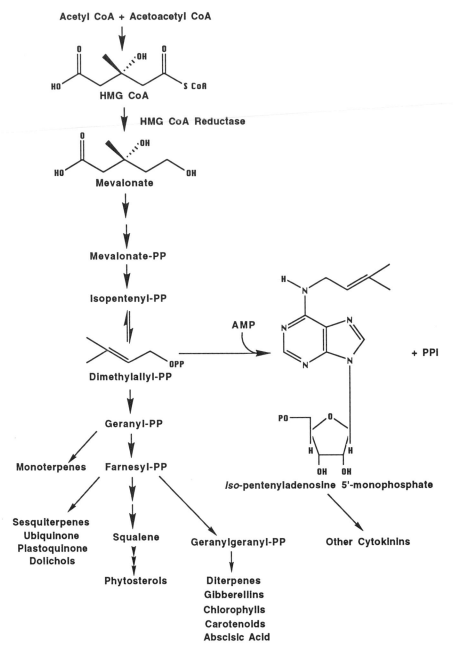

Figure 1. Biosynthesis of isoprenoids in plant cells. The cytokinin synthase shown is a dimethylallyl pyrophosphate: AMP-dimethylallyl transferase (Δ^2-isopentenyl pyrophosphate: AMP-Δ^2-isopentenyl transferase). However, other acceptor molecules for the dimethylallyl group are possible. Portions taken from the following published papers (1, 3, 5, 8, 9, 13). P, phosphate; AMP, adenosine 5'-monophosphate.

been shown to affect plant HMG CoA reductase activity) (3, 7, 8, 9, 20) and is strongly inhibited by lovastatin (mevinolin), an antibiotic isolated from *Aspergillus terreus* (2, 4, 6). We report that lovastatin inhibits the growth of a cytokinin-autonomous tobacco suspension culture at low concentrations and that this inhibition is reversed by cytokinin. These data suggest that low concentrations of lovastatin specifically inhibit cytokinin biosynthesis.

MATERIALS AND METHODS

Tissue Culture

All experiments were done on suspension cultures of *Nicotiana tabacum* cell line BY-2 (a cytokinin-independent line), which were grown at 26 ±1 °C in 18 mL of Murashige-Skoog (17) medium containing 0.2 mg/L 2, 4-dichlorophenoxyacetic acid. Cultures were started by adding 3 mL of a 2-fold diluted stationary phase culture (10 days old) to 15 mL of fresh medium and were grown in continuous white light on a rotary shaker set to 100 rpm. All additions to the cultures (e.g., lovastatin, cytokinin, etc.) were made after 24 hours to allow the cells to acclimate to fresh medium. (Day 0 is defined as the day that additions were made.) Growth was monitored every two days by measuring 10 minute settled cell volumes (clumping precluded accurate measurement of cell number) and by microscopic examination of the cells (which indicated that changes in cell size could not account for the observed changes in settled cell volume). All experiments were done in duplicate.

All phytohormones were obtained from Sigma (St. Louis, MO) except thidiazuron, which was a generous gift from Dr. Richard Amasino.

Preparation of Chemicals

Lovastatin was kindly provided by A. Alberts of Merck Sharp & Dohme Research Laboratories. A 5 mg/mL stock solution of lovastatin was prepared after hydrolyzing the lactone ring in ethanolic NaOH (15% ethanol, v/v; 0.25% NaOH, w/v) at 60°C for 1 hour. Stock solutions of mevalonic acid (50 mg/mL) were prepared by the same procedure. A 'blank' solution containing ethanol and NaOH but no lovastatin or mevalonic acid had no effect on the cells used in this study.

All cytokinins were dissolved in 1 M HCl and then diluted such that stock solutions contained 1mM HCl (pH~3). At the concentrations used in this study, these cytokinins did not affect the pH of the growth medium.

Microscopy

Cells were examined with a Leitz Dialux 20 microscope at a magnification of 12.5.

RESULTS

Lovastatin, an antibiotic isolated from *Aspergillus terreus*, is a potent and specific inhibitor of plant (4, 6) and mammalian (2, 9) HMG CoA reductases. Low concentrations of lovastatin (i.e., less than 10 µM) have been shown to decrease phytosterol synthesis in, and inhibit normal growth of, plant seedlings and cell cultures (3, 4, 5, 10, 11), suggesting that HMG CoA reductase activity is rate-limiting for phytosterol synthesis and normal plant growth (3). However, the primary mechanism of growth inhibition by lovastatin remains unclear.

This work was undertaken to identify the primary cause of growth inhibition by lovastatin and thereby achieve a better understanding of the role of isoprenoids in plant cell growth. Our strategy was to search for an isoprenoid compound (or combination of compounds) that would restore growth to lovastatin-treated tobacco cells. Accordingly, the growth response of a cytokinin-independent *Nicotiana tabacum* cell line, called BY-2, to various

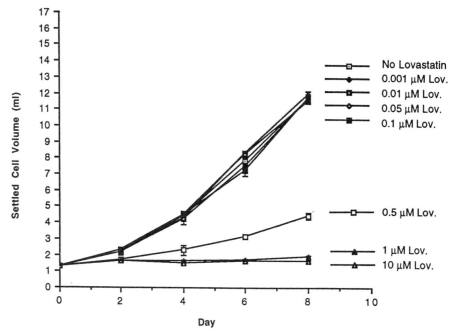

Figure 2. Lovastatin inhibits growth of cultured tobacco cells. All values represent ten minute settled cell volumes (measured in milliliters), which were recorded on Days 0, 2, 4, 6 and 8 of the experiment. Error bars represent standard deviations. Lov., Lovastatin.

concentrations of lovastatin was measured. As shown in Figure 2, 0.5 µM lovastatin partially inhibited cell growth and concentrations of lovastatin greater than or equal to 1 µM completely inhibited cell growth. Microscopic examination of the cells also indicated that inhibition of cell growth at 1 µM lovastatin correlated with loss of cell viability, as judged by staining with 0.05% Evans Blue (data not shown).

Naturally occurring cytokinins are mevalonic acid-derived compounds that regulate plant growth and development. Accordingly, we examined the effects of various cytokinins on growth of lovastatin-treated tobacco cells. As shown in Figure 3, 8 μM zeatin partially restored growth to cultures treated with 1 μM lovastatin, indicating that growth inhibition at this concentration of lovastatin was caused, at least in part, by reduced endogenous synthesis of cytokinin or other essential isoprenoid(s) that can be synthesized from zeatin. However, 8 μM benzyladenine, 8 μM kinetin and 8 μM thidiazuron also restored growth to cells treated with 1 μM lovastatin, arguing against the latter (these compounds are not isoprenoids). With the exception of kinetin, these cytokinins were not effective at lovastatin concentrations above 1 μM, suggesting that lovastatin either inhibits the biosynthesis of other essential isoprenoids or has non-specific, toxic effects at high concentrations (data not shown). However, 6 mM mevalonic acid was found to restore growth at all lovastatin concentrations tested (≤ 40 μM), arguing against non-specific cytotoxicity at high lovastatin concentrations. Together, these observations suggest that 1 μM lovastatin specifically inhibited cytokinin biosynthesis, whereas higher lovastatin concentrations also inhibited the synthesis of other essential isoprenoids. Interestingly, 8 μM kinetin restored growth to cells treated with concentrations of lovastatin ≤ 20 μM, but had a mild inhibitory effect on growth in the absence of lovastatin (data not shown).

To define the specificity of the cytokinin effect described above, adenine and 6-methyladenine (both cytokinin analogs with little or no cytokinin activity) were tested for their ability to reverse the inhibitory effects of lovastatin on cell growth (14). As shown in Figure 3, 8 μM adenine and 8 μM 6-methyladenine did not restore growth to cells treated with 1 μM lovastatin, indicating that this phenomenon is cytokinin-specific.

DISCUSSION

The results described in this paper suggest that cytokinin biosynthesis is more sensitive to lovastatin than the biosynthesis of other essential isoprenoids in tobacco cells. Hence, lovastatin can be used at low concentrations to induce cytokinin dependence in cytokinin-independent cell cultures and is, thus, a potentially useful tool for altering cytokinin levels *in vivo*.

Interestingly, other researchers have proposed that lovastatin might cause imbalances in phytohormone levels (5). In an earlier study of the effects of compactin (an analog of lovastatin) on growth of cytokinin-dependent tobacco callus, it was found that kinetin lessened the inhibitory effect of 5 μM compactin by 1.4-fold (10). In contrast, we find that for a cytokinin-independent tobacco culture, kinetin relieves the complete inhibition of growth imposed by lovastatin and, thus, has an stimulatory effect on growth in the presence of lovastatin greater than 10-fold.

The apparent specificity of lovastatin action at low concentrations can be explained in at least two ways. First, it is possible that multiple HMG CoA reductases exist in the plant cell that are dedicated, perhaps by intracellular compartmentalization, to different aspects of isoprenoid metabolism. If these HMG CoA reductases differ in their sensitivity to lovastatin, low lovastatin concentrations might thereby exert specific effects on isoprenoid biosynthesis.

Second, it is possible that the many enzymes involved in plant isoprenoid metabolism have different affinities for their respective substrates. If the dimethylallyl pyrophosphate:AMP-dimethylallyl transferase has a low affinity (i.e., a high K_m) for dimethylallyl pyrophosphate, then low concentrations of lovastatin might deplete the pool of isoprenoid precursors such that cytokinin biosynthesis is preferentially inhibited. These models predict that cytokinins will not restore growth at high lovastatin concentrations, but mevalonic acid will.

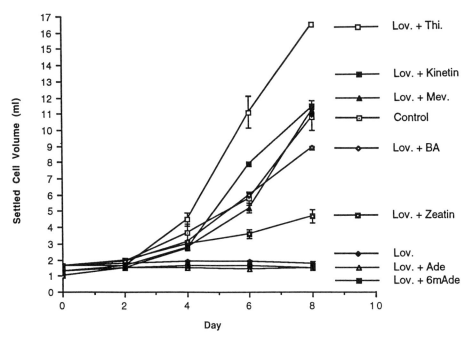

Figure 3. 8 μM zeatin, 8 μM benzyladenine, 8 μM kinetin, 8 μM thidiazuron and 6 mM mevalonic acid restore growth to cells treated with 1 μM lovastatin, but 8 μM adenine and 8 μM 6-methyladenine do not. All values represent ten minute settled cell volumes (measured in milliliters), which were recorded on Days 0, 2, 4, 6 and 8 of the experiment. Error bars represent standard deviations. Lov., Lovastatin; BA, benzyladenine; Thi., thidiazuron; Mev., Mevalonate; Ade, adenine; 6mAde, 6-methyladenine.

Consistent with the two models described above, we observed that zeatin, benzyladenine and thidiazuron (but not mevalonic acid) lose effectiveness at lovastatin concentrations greater than 1 μM. However, kinetin restores growth to cells treated with higher concentrations of lovastatin ($\leq$ 20 μM), an observation for which various explanations can be proposed. It is possible, for example, that lovastatin induces cytokinin turnover at concentrations above 1 μM and that kinetin is more resistant to degradation than the other cytokinins tested. Alternatively, it can be argued that total cytokinin levels (endogenous and exogenous) fall below a critical threshold above 1 µM lovastatin except in the case of exogenously added kinetin (which may be more stable than the other exogenously added cytokinins). Finally, these cytokinins may directly or indirectly compete with lovastatin, either at the level of HMG CoA reductase or at the level of entry into the cell, and kinetin may be a more effective competitor than the other cytokinins tested. An explanation for this complicated phenomenology will require a rigorous biochemical analysis of lovastatin-induced changes in cytokinin content, sterol content, and HMG CoA reductase activity in the presence and absence of exogenous cytokinin.

The data presented in this paper provide further evidence of the relationship between the mevalonate pathway and cytokinin biosynthesis in plant cells (1, 13). However, reduction of HMG CoA to mevalonate is probably not the rate-limiting step in cytokinin biosynthesis. Rather, dimethylallyl pyrophosphate: AMP-dimethylallyl transferase activity is likely to be rate-limiting since overexpression of this activity results in cytokinin overproduction and tumor formation (a potentially misleading result if the enzyme from *Agrobacterium tumefaciens* utilizes stores of dimethylallyl pyrophosphate that are unavailable to the native cytokinin synthase) (1). This observation suggests that the isoprenoid precursor to cytokinin, dimethylallyl pyrophosphate, is present in an excess in plant cells. The data in this paper also provide evidence that cytokinin-autonomous cell lines require cytokinin for growth, a conclusion based on the observation that inhibition of isoprenoid biosynthesis inhibits growth and that cytokinins restore growth. This conclusion lends support to the hypothesis that cytokinins are universally required for cell division in plants.

ACKNOWLEDGMENTS

The authors wish to thank Drs. Martin Bard, Stephen Randall, Richard Amasino and Pamela Crowell for technical advice and for critical reading of the manuscript. This work was supported by the Purdue School of Science at Indianapolis and by an IUPUI faculty development grant.

REFERENCES

1. Akiyoshi DE, Klee H, Amasino RM, Nester EW, Gordon MP (1984) T-DNA of *Agrobacterium tumefaciens* encodes an enzyme of cytokinin biosynthesis. Proc Natl Acad Sci USA **81**: 5994-5998
2. Alberts AW, Chen J, Kuron G, Hunt V, Huff J, Hoffman C, Rothrock J, Lopez M, Joshua H, Harris E, Patchett A, Monagan R, Currie S, Stapley E, Albers-Schönberg G, Hensens O, Hirschfield J, Hoogsteen K, Liesch J, Springer J (1980) Mevinolin: a highly potent competitive inhibitor of hydroxymethylglutaryl-coenzyme A reductase and a cholesterol-lowering agent. Proc Natl Acad Sci USA **77**: 3957-3961
3. Bach TJ (1986) Hydroxymethylglutaryl-CoA reductase, a key enzyme in phytosterol synthesis? Lipids **21**: 82-88
4. Bach TJ, Lichtenthaler HK (1982) Mevinolin: a highly specific inhibitor of microsomal 3-hydroxy-3-methylglutaryl-coenzyme A reductase of radish plants. Z. Naturforsch **37c**: 46-50

5. Bach TJ, Lichtenthaler HK (1983) Inhibition by mevinolin of plant growth, sterol formation and pigment accumulation. Physiol Plant **59**: 50-60
6. Bach TF, Lichtenthaler HK (1983) Mechanisms of inhibition by mevinolin (MK803) of microsome-bound radish and of partially purified yeast HMG-CoA reductase (EC.1.1.1.34). Z. Naturforsch **38c**: 212-219
7. Brooker JD, Russell DW (1979) Regulation of microsomal 3-hydroxy-3-methylglutaryl coenzyme A reductase from pea seedlings: rapid posttranslational phytochrome-mediated decrease in activity and *in vivo* regulation by isoprenoid products. Arch Biochem Biophys **198**: 323-334
8. Brown MS, Goldstein JL (1980) Multivalent feedback regulation of HMG CoA reductase, a control mechanism coordinating isoprenoid synthesis and cell growth. J Lipid Res **21**: 505-517
9. Goldstein JL, Brown MS (1990) Regulation of the mevalonate pathway. Nature **343**: 425-430
10. Hashizume T, Matsubara S, Endo A (1983) Compactin (ML-236B) as a new growth inhibitor of plant callus. Agric Biol Chem **47**: 1401-1403
11. Hata S, Takagishi H, Kouchi H (1987) Variation in the content and composition of sterols in alfalfa seedlings treated with compactin (ML-236B) and mevalonic acid. Plant Cell Physiol **28**: 709-714
12. Letham DS, Higgins TJV, Goodwin PB, Jacobsen JV (1978) Phytohormones in retrospect. *In* DS Letham, PB Goodwin, TJV Higgins, eds, Phytohormones and Related Compounds: A Comprehensive Treatise, Vol 1. Elsevier/North-Holland, New York, pp 1-27
13. Letham DS, Palni LMS (1983) The biosynthesis and metabolism of cytokinins. Ann Rev Plant Physiol **34**: 163-197
14. Matsubara S (1980) Structure-activity relationships of cytokinins. Phytochemistry **19**: 2239-2253
15. Miller CO (1968) Biochemistry and Physiology of Plant Growth Substances. Runge, Ottawa, pp 33-45
16. Miller CO, Skoog F, Von Saltza MH, Strong FM (1955) Kinetin, a cell division factor from deoxyribonucoeic acid. J Am Chem Soc **78**: 1392
17. Murashige T, Skoog F (1962) A revised medium for rapid growth and bio assays with tobacco tissue cultures. Physiol Plant **15**: 473-497
18. Noodén LD, Guiamét JJ, Singh S, Letham DS, Tsuji J, Schneider MJ (1990) Hormonal control of senescence. *In* RP Pharis, SB Rood, eds, Plant Growth Substances. Springer-Verlag, Berlin, pp 537-546
19. Richmond AE, Lang A (1957) Effect of kinetin on protein content and survival of detached *Xanthium* leaves. Science **125**: 650-651
20. Russell DW, Davidson H (1982) Regulation of cytosolic HMG-CoA reductase activity in pea seedlings: contrasting responses to different hormones, and hormone-product interaction, suggest hormonal modulation of activity. Biochem Biophys Res Comm **104**: 1537-1543

MOLECULAR GENETIC APPROACHES TO ELUCIDATING THE ROLE OF HORMONES IN PLANT DEVELOPMENT

Harry Klee and Charles Romano

Monsanto Company
700 Chesterfield Village Parkway
Chesterfield, MO 63198

INTRODUCTION

Physiological studies conducted over the last half century have established a role for hormones in virtually every aspect of plant development. Most of the pioneering work on hormone action used approaches that involve exogenous application of a hormone or inhibitor. There are severe limitations to what we can learn in this manner. Exogenous application of any biological material is subject to limitations of uptake, transport, sequestration and metabolism. Further, it is difficult to quantitate the amount of active material within the target tissue. For these reasons, it has been generally difficult to establish a direct relationship between a hormone and a particular developmental process.

There are many examples of interactions between hormones in the literature. For example, it is clear from tissue culture studies (Skoog and Miller, 1957), as well as the experiments with transgenic plants discussed below, that the ratio of auxin to cytokinin is a critical factor in many developmental processes. The phenotype of several hormone-resistant mutants of *Arabidopsis* suggest that some signal transduction components may be shared by several hormones. The axr mutants of *Arabidopsis* were originally isolated by screening for resistance to auxin (Estelle and Somerville, 1987). Subsequently, these mutants were found to be resistant to at least one other plant hormone, indicating that a single mutation can disrupt response to several hormones. For example, the dominant axr2 mutant has roots that are resistant to the growth inhibiting effects of ethylene, ABA, and cytokinin, in addition to auxin (Wilson et al., 1990). Another type of hormone interaction involves regulation of hormone levels by a second hormone. For example, auxin has been shown to stimulate the synthesis of ethylene in many plant tissues (Yang and Hoffman, 1984).

The development of molecular approaches to studying plant hormone action has been accompanied by an increase in the use of mutants to study hormone processes in plants (King, 1988; Reid, 1990). Mutants that are either hormone deficient or deficient in one or more hormone responses have been identified. In addition, genes that can alter the levels of the various hormones have also become available. Having a gene that directs synthesis of a hormone or a rate-limiting precursor opens up the possibility of increasing its level in a controlled manner in vivo. Availability, in some cases, of the plant genes also permits their shut-off by using antisense RNA techniques. Thus, a hormone can be increased or decreased in very precise and reproducible ways. Alteration of the steady-state level of a hormone permits a direct test of the biological effects of the hormone on a tissue. As more genes become available, there is great potential for increasing our knowledge of hormone synthesis and action, permitting us to manipulate plant growth and development for agronomic and horticultural purposes.

Genes and mutants controlling hormone metabolism and action

iaaM and iaaH. *Agrobacterium tumefaciens* is known to cause crown gall disease because it encodes genes that direct auxin and cytokinin synthesis. IAA is synthesized in a two step process in crown gall tumors. The first gene, tryptophan monooxygenase (iaaM), converts tryptophan to indole-3-acetamide (IAM). This intermediate is then converted to IAA by indoleacetamide hydrolase (iaaH). An increase in the level of IAA can be achieved by overexpressing iaaM alone if a strong transcriptional promoter such as the cauliflower mosaic virus (CaMV) 19S or 35S promoter is used (Klee et al.1987). In this circumstance, conversion of IAM to IAA must occur via hydrolysis, either chemically or by a non-specific amidohydrolase. Expression of a 19S/iaaM gene in petunia or tobacco can lead to a 10-fold increase in free IAA. Generally, the increase in auxin is accompanied by a 3-5-fold increase in ethylene production as well.

ipt. The product of the *Agrobacterium* cytokinin synthesis gene, isopentenyl transferase (ipt), condenses isopentenyl pyrophosphate and AMP to produce isopentenyl AMP (iPMP). Synthesis of iPMP is probably the rate limiting reaction in cytokinin biosynthesis since it is rapidly converted by the plant to a series of more biologically active cytokinins, most notably zeatin derivatives (Medford et al.1989). Attempts to introduce this gene into plants have proven difficult. Even under the control of weak transcriptional promoters, transgenic shoots do not form roots. Fertile transgenic plants have been obtained when the gene is placed under the control of regulated promoters such as the hsp70 or tissue specific promoters (Medford et al.1989). Transgenic tobacco plants containing an hsp70/ipt gene have increased cytokinins even in the absence of heat shock. Heat shock results in increases of as much as 100-fold in different cytokinins.

IAA-Lysine synthetase. Another example of an organism that has provided useful genes is *Pseudomonas syringae* pv. *savastanoi*. Besides being able to synthesize IAA *P. savastanoi* contains a gene (iaaL) that conjugates IAA to lysine (Glass and Kosuge, 1986; Roberto et al.1990). Auxin-amino acid conjugates are much less biologically active and are believed to be storage forms of the hormone (Cohen and Bandurski, 1982). Constitutive overproduction of iaaL in tobacco, under the control of the CaMV 35S promoter, leads to upwards of twenty-fold reduction in free IAA and concomitant increases in IAA-lysine (Romano et al. 1991). The transgenic plants exhibit morphological effects suggestive of auxin deprivation.

ACC deaminase. Until recently, all of the progress in isolating phytohormone metabolic genes has been limited to bacteria. Now, many groups have isolated genes involved in ethylene synthesis. Ethylene is synthesized from S-adenosylmethionine via the intermediate, 1-aminocyclopropane-1-carboxylic acid (ACC). We took an alternative strategy for controlling ethylene synthesis. A bacterial gene encoding an ACC deaminase was cloned and expressed in transgenic plants. Plants expressing ACC deaminase are reduced by up to 97% in the level of ethylene synthesized. These transgenic plants are therefore, essentially ethylene auxotrophic.

ein mutants. An additional tool that we have used to dissect out the relative roles of auxin and ethylene are the ethylene insensitive mutants of *Arabidopsis*. Several of these have now been described (Bleecker, et al., 1988; Guzman and Ecker, 1990). Because the etr and ein1-1 mutation are both dominant and map to the same chromosomal region, they are probably allelic. The ein2 mutation is recessive and unlinked to etr and ein1. A wide range of ethylene responses have been characterized in these mutants, including inhibition of hypocotyl elongation, stimulation of germination, induction of chlorophyll loss, and accumulation of peroxidases (Bleecker et al., 1988). Generally, the mutants are completely insensitive to ethylene.

RESULTS

Perhaps the most remarkable aspect of introduction of genes that alter hormone concentrations is that fertile plants can be recovered. In tobacco the range of IAA concentration between overproducers (iaaM) and underproducers (iaaL) is two hundred-fold. With cytokinins, the picture is a little more complicated because of the multiple forms of cytokinins, but increases of 100-200-fold are tolerated. While there are a number of major effects on different developmental processes, it is fair to say that a large number of cells and organs do not respond to alterations in hormone levels. For example, overproduction of auxin or cytokinin in maturing petunia embryos has no detectable effect ((Klee et al.1987) and H. Klee, unpublished). This type of result indicates that perception, as well as synthesis of hormones is developmentally controlled.

One important conclusion of the transgenic plant experiments is that there is a complex interaction between cytokinins, auxins and ethylene that cannot be easily separated out. For example, overproduction of IAA with the A. tumefaciens genes has some major phenotypic effects on plants (discussed below). Significant increases in auxin stimulate ethylene production and it has been difficult to separate the "auxin" effects from ethylene effects. For this reason, we have set up systems which allow us to examine auxin effects in the absence of ethylene effects (Romano, C. and Klee, H., submitted for publication). This goal has been accomplished in two ways. First, we have produced transgenic plants that express ACC deaminase and as a consequence, have reduced ethylene production. When such plants are crossed to plants overexpressing the auxin biosynthetic gene, the F1 hybrids contain greatly elevated auxin but no increase in ethylene over controls. The second way we have separated auxin and ethylene effects is to cross auxin overproducing *Arabidopsis* plants with the ethylene insensitive ein1 and ein2 mutants. In both cases, the results are the same, confirming the validity of the two approaches.

That auxin and cytokinin are capable of controlling apical dominance is well established (Tamas, 1987). High auxin suppresses release of lateral buds from dormancy while cytokinin stimulates their growth. Thus, it is not surprising that plants overexpressing the iaaM gene exhibit almost complete apical dominance (Klee et al.1987) and plants overexpressing the ipt gene exhibit reduced apical dominance (Smigocki and Owens, 1989; Medford et al.1989). What is more surprising is that the extreme apical dominance in auxin-overproducing plants can be overcome by either crossing them with cytokinin overproducing plants or exogenously applying cytokinin to a dormant lateral bud (H. Klee, unpublished). These results suggest that absolute levels of auxin and cytokinin are not the determinants of lateral growth. Rather, it is the ratio of auxin to cytokinin that controls growth. Dormancy can be induced by raising the auxin level ten-fold but can be relieved by further increasing the cytokinin in the bud. If this is the case, then reducing the effective auxin level should be equivalent to increasing the cytokinin level. This prediction was verified when plants overexpressing the iaaL gene were produced (Romano et al. 1991). These plants, containing 5- to 20-fold lower IAA levels than controls, exhibited a reduced apical dominance similar to the ipt overproducing plants. One additional question concerns the role of auxin in reducing lateral bud growth. It has been suggested that auxin acts by stimulating high ethylene production and that it is actually ethylene that is the controlling hormone. Our experiments clearly rule out this possibility. There is no difference in the degree of apical dominance in an auxin overproducing plant with or without increased ethylene or in an ethylene insensitive genetic background. These results have been confirmed in tobacco, petunia and *Arabidopsis*. Thus, lateral growth appears to be regulated by the ratio of these two hormones and absolute levels appear to be secondary.

The effect of auxin on vascular differentiation is well established (Aloni, 1988). Auxin is believed to affect xylem formation both quantitatively and qualitatively. Work with transgenic plants having altered levels of IAA generally confirm the direct relationship between auxin and the degree of differentiation of xylem ((Klee et al.1987; Romano et al, 1991) . Auxin overproducing plants contain more xylem elements than control plants. The cells are, however, smaller. Conversely, plants with lowered IAA levels contain fewer xylem elements and they are generally of a larger size. These results support the idea that auxin stimulates cell division within the vascular cambium and that there is a direct relationship between auxin content and rate of cell division(Aloni and Zimmermann, 1983). Cell size is most likely affected because auxin stimulates secondary cell wall formation. The more rapidly the wall is synthesized, the less time there is for cell expansion.

Both auxin and cytokinin perturbations have other interesting effects on plant growth. Auxin alteration can have profound effects on leaf development. Overproduction can lead to epinastic growth. Generally leaves are smaller and narrower than controls. The epinasty results from uneven expansion of cell layers with the upper layers showing greater expansion than the lower layers. Initially, we anticipated that this epinastic growth was a consequence of induced ethylene synthesis. The experiments with ACC deaminase and ein plants clearly indicate that the epinasty is an auxin effect and is unrelated to ethylene.

Underproduction of auxin leads to a very dramatic wrinkling of the leaf. Histological analysis suggests that the wrinkling is due to incomplete development of the vascular system relative to the rest of the leaf.

CONCLUSIONS

The classic experiments of Skoog and Miller (1957) established a precedent for an antagonistic relationship between auxin and cytokinin. They demonstrated that undifferentiated tobacco tissue could be induced to form roots, shoots or undifferentiated callus depending on the ratio of auxin to cytokinin. Analysis of hormone mutants and transgenic plants with altered auxin and cytokinin levels has confirmed the complex interactions between hormones. Many processes such as apical dominance are relatively insensitive to absolute levels of hormone and seem to be regulated by the auxin to cytokinin ratio. There is further an intimate relationship between auxin and ethylene. This is similar to the situation in seed dormancy and germination which appears to be regulated by the antagonistic balance of ABA and GA. This interrelationship of the hormones suggests that they may act antagonistically through common signal transduction pathways. The metabolic and/or developmental fate of the cell is then determined by the degree of stimulation of that transduction pathway. More auxin would be equivalent to less cytokinin and vice versa. An alternative interpretation that cannot be ruled out is that one hormone may act by altering either the level or sensitivity to a second hormone. It is known, for example, that auxin can influence the stability and metabolism of cytokinin (Palni et al.1988). Thus, it will be critical to measure the levels of all hormones in mutants and transgenic plants to obtain a complete picture.

It is likely that many of the genes encoding additional hormone biosynthetic activities will be cloned in the next several years. We are thus likely to have available to us tools for manipulation of all of the phytohormones within the next 5-10 years, if not sooner. These tools will be very useful both in elucidating the biological roles of the molecules and in allowing for manipulation of their levels in ways that will be beneficial to agriculture.

REFERENCES

Aloni, R. 1988. Vascular differentiation within the plant. In Vascular differentiation and plant growth regulators, 39-62, ed. T.E. Timell. Berlin Heidelberg: Springer-Verlag.

Aloni, R., Zimmermann, M.H. 1983. The control of vessel size and density along the plant axis-a new hypothesis. Differentiation 24:203-208.

Bleecker, A.B., Estelle, M.A., Somerville, C. Kende, H. (1988) Insensitivity to ethylene conferred by a dominant mutation in *Arabidopsis thaliana*. Science 241:1086-1089.

Cohen, J.D., Bandurski, R.S. 1982. Chemistry and Physiology of the Bound Auxins. Ann. Rev. Plant Physiol 33:403-430.

Estelle, M.A., Somerville, C.R. (1987) Auxin-resistant mutants of *Arabidopsis* with an altered morphology. Mol. Gen. Genet. 206:200-206.

Glass, N.L., Kosuge, T. 1986. Cloning of the gene for indoleacetic acid-lysine synthetase from *Pseudomonas syringae* subsp. *savastanoi*. J.Bacteriol. 166:598-602

Guzman, P., Ecker, J.R. (1990) Exploiting the triple response of *Arabidopsis* to identify ethylene-related mutants.

King, P.J. (1988) Plant hormone mutants. Trends Genet. 4:157-162.

Klee, H.J., Hayford, M.B., Kretzmer, K.A., Barry, G.F., Kishore, G.M. 1991. Control of ethylene synthesis by expression of a bacterial enzyme in transgenic tomato plants. Pl..Cell 3:1187-1193.

Klee, H.J., Horsch, R.B., Hinchee, M.A., Hein, M.B., Hoffmann, N.L. 1987. The effects of overproduction of two *Agrobacterium tumefaciens* T-DNA auxin biosynthetic gene products in transgenic petunia plants. Genes & Dev. 1:86-96.

Medford, J.I., Horgan, R., El-Sawi, Z., Klee, H.J. 1989. Alterations of endogenous cytokinins in transgenic plants using a chimeric isopentenyl transferase gene. T. Pl. Cell 4:403-413.

Palni, L.M., Burch, L., Horgan, R. 1988. The effect of auxin concentration on cytokinin stability and metabolism. Planta 174:231-234.

Reid, J.B. (1990) Phytohormone mutants in plant research. J. Plant Growth Regul. 9:97-111.

Roberto, F.F., Klee, H., White, F., Nordeen, R., Kosuge, T. 1990. Expression and fine structure of the gene encoding indole-3-acetyl-L-lysine synthetase from *Pseudomonas savastanoi*. Proc.Natl.Acad.Sci.USA 87:5797-5801.

Romano, C., Hein, M. and Klee, H. 1991. Inactivation of auxin in tobacco transformed with the indoleacetic acid-lysine synthetase gene of *Pseudomonas savastanoi*. Genes & Dev. 5:438-446.

Skoog, F., Miller, C.O. 1957. Chemical regulation of growth and organ formation in plant tissues cultured in vitro. Symp. Soc. Exptl. Biol. 11:188-231.

Smigocki, A., Owens, L. 1989. Cytokinin-to-auxin ratios and morphology of shoots and tissues transformed by a chimeric isopentenyl transferase gene. Plant Physiol. 91:808-811.

Tamas, I.A. 1987. Hormonal regulation of apical dominance. In Plant hormones and their roles in plant growth and development, 393-410, ed. P.J. Davies. Boston: Martinus Nijhoff Publishers.

Wilson, A.K., Pickett, F.B., Turner, J.C. Estelle, M. (1990) A dominant mutation in *Arabidopsis* confers resistance to auxin, ethylene and abscisic acid. Mol. Gen. Genet. 222:377-383.

Yang, S.F. Hoffman, N.E. (1984) Ethylene biosynthesis and its regulation in higher plants. Ann. Rev. Plant Physiol. 35:155-189.

REVERSIBLE INHIBITION OF TOMATO FRUIT RIPENING BY ANTISENSE ACC SYNTHASE RNA

Athanasios Theologis, Paul W. Oeller and Lu Min-Wong

USDA/U.C. Berkeley Plant Gene Expression Center
800 Buchanan Street
Albany, CA 94706

INTRODUCTION

Ethylene is one of the simplest organic molecules with biological activity. Its effects on plant tissue are spectacular and commercially important (1,2). This hydrocarbon gas is generally considered to be the fruit ripening hormone (2,3). Because of its effects on plant senescence, large losses of fruits and vegetables are incurred annually in the U.S. The losses are much greater in third world countries because of the lack of sufficient refrigeration and transportation. Consequently, it has always been a goal of plant biologists and of postharvest physiologists, in particular, to be able to prevent or delay fruit ripening in a reversible manner by controlling ethylene action or production. Thus, an understanding of ethylene action and biosynthesis is of fundamental as well as of applied significance. This update summarizes the recent advances in manipulating key genes in the ethylene biosynthetic pathway to prevent ethylene production and fruit ripening.

Methionine is the biological precursor of ethylene in all higher plants (22), which is converted to ethylene according to the following sequence:

$$\text{Methionine} \xrightarrow{1} \text{AdoMet} \xrightarrow{2} \text{ACC} \xrightarrow{3} \text{Ethylene.}$$

The rate limiting step in the pathway is the formation of the amino acid ACC from AdoMet, catalyzed by ACC synthase (reaction 2). The final step is the conversion of ACC to C_2H_4 catalyzed by ACC oxidase (reaction 3). Molecular cloning approaches and expression in heterologous systems allowed the isolation of the genes encoding AdoMet synthase (step 1, [14]), ACC synthase (step 2, [11,16,21]), and ACC oxidase (step 3, [6,7,18,19]).

Abbreviations: AdoMet, S-adenosylmethionine; ACC, 1-aminocyclopropane-1-carboxylic acid; PG, polygalacturonase.

INHIBITION OF FRUIT RIPENING USING REVERSE GENETICS

Ethylene is thought to regulate fruit ripening by coordinating the expression of genes that are responsible for a variety of processes, including enhancing a rise in the rate of respiration, autocatalytic ethylene production, chlorophyll degradation, carotenoid synthesis, conversion of starch to sugars, and increased activity of cell wall degrading enzymes (5). Throughout the years, various methods for prolonging fruit senescence have been employed, such as ventilation with air under hypobaric pressures (4). The procedure accelerates the escape of ethylene and, by reducing the oxygen tension, also lowers the fruits' sensitivity to hormone. Also, inhibitors of ethylene action have been used such as silver ions (Ag^+) and carbon dioxide (CO_2) (22). These approaches are expensive and fail to prevent fruit senescence satisfactorily. In a few cases, however, such as apple, controlled atmosphere storage has been a commercial success. A more desirable solution to the problem will be the construction of a mutant plant whose fruits do not ripen unless they are treated with ethylene. Tomato ripening mutants exist, but their phenotype is not reversible by ethylene (10).

The cloning of genes induced during fruit ripening and of genes involved in ethylene biosynthesis opened the road to the construction of ripening mutants in tomato using reverse genetics. In the absence of gene replacement technology in plants, antisense RNA technology and overexpression of an ACC metabolizing enzyme became the tools of choice (5).

Antisense RNA

Initially, attempts to inhibit tomato fruit softening by antisense polygalacturonase (PG) RNA, a gene thought to be responsible for cell wall hydrolysis during ripening, failed to give a strong effect (17,18). Expression of PG antisense RNA dramatically inhibited PG mRNA accumulation and enzyme activity, suggesting that PG is not the sole determinant of cell wall hydrolysis (18). Another approach to prevent fruit ripening is to inhibit ethylene production. Hamilton et al. (6) inhibited ACC oxidase activity with antisense RNA. In plants that were homozygous for the antisense gene, ethylene production was inhibited by 97% in ripening fruit. In these antisense fruits, the color change was initiated at about the normal time; however, the extent of reddening was reduced. Antisense fruits stored for several weeks at room temperature were more resistant to overripening and shrivelling than control fruits (6). More recently, Oeller et al. (12) used antisense RNA to ACC synthase to inhibit tomato fruit ripening. This approach led to severe inhibition of ethylene production (below 0.1 nl/g·h; 99% inhibition) resulting in a tomato fruit mutant with a striking phenotype (Figure 1). This dramatic inhibition of ethylene production can be attributed to the short half life of ACC synthase (8). Antisense experiments are intrinsically "leaky", thus allowing some mRNA to be translated. Consequently, the stability of the encoded polypeptide is an important factor for successful gene inactivation by antisense RNA (12).

During tomato fruit ripening, two ACC synthase genes are expressed, LE-ACC2 and LE-ACC4 (13,15). Expression of antisense RNA derived from the cDNA of the LE-ACC2 gene resulted in an almost complete inhibition of mRNA accumulation of both ripening-induced ACC synthase genes (12). Control fruits kept in air begin to produce ethylene 50 days after pollination and fully ripen after 10 more days. The red coloration resulting from chlorophyll degradation and lycopene biosynthesis is inhibited in antisense fruits. Antisense fruits kept in air or on the plants for 90 to 150 days eventually develop an orange color but never turn red and soft or develop an aroma.

The antisense phenotype can be reversed by treatment with ethylene or propylene, an ethylene analog. The treated fruits are *indistinguishable* from naturally ripened fruits with respect to texture, color, aroma, and compressibility. The duration of C_2H_4 treatment required to reverse the antisense phenotype is 6 days. Antisense fruits treated for 1 or 2 days with C_2H_4 do not develop a fully ripe phenotype compared with control fruits treated similarly.

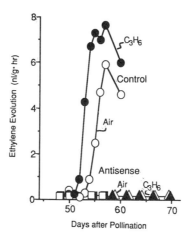

Figure 1. Inhibition of ethylene production in detached tomato fruits by antisense ACC synthase RNA (12). Reproduced with permission of *Science*.

ACC Deaminase

Klee et al. (9) used a different approach to inhibit ethylene production. They overexpressed the ACC deaminase gene from *Pseudomonas* sp. in transgenic tomato plants which metabolizes ACC to α-ketobutyrate (9). This approach led to 90-97% inhibition in ethylene production during ripening. Reduction in ethylene synthesis in transgenic plants did not cause any apparent vegetative phenotypic abnormaliities. However, fruits from these plants showed significant delays in ripening, and they remained firm for at least 6 weeks longer than the nontransgenic control fruits (9).

Three important conclusions can be inferred from the tomato fruit mutants: (i) The C_2H_4-mediated ripening process requires continuous transcription of the necessary genes, which may reflect a short half-life of the induced mRNAs or polypeptides. (ii) Ethylene is indeed autocatalytically regulated. (iii) The hormone acts as a rheostat rather than as a switch for controlling the ripening process. The accumulated evidence from the antisense and the deaminase experiments indisputably demonstrated that the Yang cycle is solely responsible for ethylene synthesis during ripening and that ethylene is the key regulatory molecule for fruit ripening and senescence, not the by-product of ripening (2).

The mutant fruits producing low levels of ethylene have proven to be excellent experimental material for assessing which of the ripening induced genes so far cloned are indeed ethylene inducible. The expression of PG and pTOM13 (encodes ACC oxidase) genes, which were previously thought to be ethylene regulated, have been found to be ethylene independent (12). Surprisingly, while antisense fruits express large amounts of PG mRNA, they fail to accumulate the PG polypeptide, suggesting that ethylene may control the translatability of PG mRNA or the stability of the PG polypeptide (20).

The mutants also told us that at least two signal transduction pathways are operating during tomato fruit ripening. The ethylene independent (developmental) pathway is responsible for the transcriptional activation of genes such as PG, ACC oxidase, and chlorophyllase. The ethylene dependent pathway, on the other hand, is responsible for the transcriptional and posttranscriptional regulation of genes involved in lycopene and aroma biosynthesis, respiratory metabolism, ACC synthase gene expression, and translatability of developmentally regulated genes such as PG (20).

THE FUTURE

The use of antisense technology and overexpression of metabolizing enzymes such as ACC deaminase in controlling fruit ripening is only the first step towards controlling fruit senescence. Expression of antisense RNA using regulated promoters may eliminate the use of exogenous ethylene for reverting the mutant phenotype. However, the development of gene transplacement technology by homologous recombination should allow the creation of non-leaky ripening mutants with long term storage potential. The prospect arises that inhibition of ethylene production using reverse genetics may be a general method for preventing senescence in a variety of fruits and vegetables.

ACKNOWLEDGMENTS

This work was supported by grants to A.T. from the NSF (DCB-8645952, -8819129, -8916286), the NIH (GM-35447) and the USDA (5835-28410-D002).

REFERENCES

1. Abeles, F. B. (1973) "Ethylene in Plant Biology", Academic Press, New York.
2. Biale, J. B. and Young, R. E. (1981) Respiration and Ripening in Fruits-- Retrospect and prospect. In: J. Friend, M. J. C. Rhodes, eds., "Recent Advances in the Biochemistry of Fruits and Vegetables", Academic Press, London, pp. 1-39.
3. Burg, S. P. (1962) The physiology of ethylene formation. *Ann. Rev. Plant Physiol.* **13**:265-302.
4. Burg, S. P. and Burg, E. A. (1966) Fruit storage at subatmospheric pressures. *Science* **153**:314-315.
5. Gray, J., Picton, S., Shabbeer, J., Schuch, W. and Grierson, D. (1992) Molecular biology of fruit ripening and its manipulation with antisense genes. *Plant Mol. Biol.* **19**:69-87.
6. Hamilton, A. J., Lycett, G. W. and Grierson, D. (1990) Antisense gene that inhibits synthesis of ethylene in transgenic plants. *Nature* **346**:284-287.
7. Hamilton, A. J., Bouzayen, M. and Grierson, D. (1991) Identification of a tomato gene for the ethylene-forming enzyme by expression in yeast. *Proc. Natl. Acad. Sci. USA* **88**:7434-7437.
8. Kende, H. and Boller, T. (1981) Wound ethylene and 1-aminocyclopropane-1-carboxylate synthase in ripening tomato fruit. *Planta* **151**:476-481.
9. Klee, H. J., Hayford, M. B., Kretzmer, K. A., Barry, G. F., Kishmore, G. M. (1991) Control of ethylene synthesis by expression of a bacterial enzyme in transgenic tomato plants. *Plant Cell* **3**:1187-1193.
10. McGlasson, W. B. (1985) Ethylene and fruit ripening. *Hort. Sci.* **20**:51-54.

11. Nakajima, N., Mori, H., Yamazaki, K. and Imaseki, H. (1990) Molecular cloning and sequence of a complementary DNA encoding 1-aminocyclopropane-1-carboxylate synthase induced by tissue wounding. *Plant Cell Physiol.* **31**:1021-1029.
12. Oeller, P. W., Wong, L.-M., Taylor, L. P., Pike, D. A. and Theologis, A. (1991) Reversible inhibition of tomato fruit senescence by antisense RNA. *Science* **254**:437-439.
13. Olson, D. C., White, J. A., Edelman, L., Harkins, R. N. and Kende, H. (1991) Differential expression of two genes for 1-aminocyclopropane-1-carboxylate synthase in tomato fruits. *Proc. Natl. Acad. Sci. USA* **88**:5340-5344.
14. Peleman, J., Boerjan, W., Engler, G., Seurinck, J., Botterman, J., Alliote, T., Van Montagu, M. and Inze, D. (1989) Strong cellular preference in the expression of a housekeeping gene of *Arabidopsis thaliana* encoding S-adenosylmethionine synthetase. *Plant Cell* **1**:81-93.
15. Rottmann, W. E., Peter, G. F., Oeller, P. W., Keller, J. A., Shen, N. F., Nagy, B., Taylor, L. P., Campbell, A. D. and Theologis, A. (1991) 1-aminocyclopropane-1-carboxylate synthase in tomato is encoded by a multigene family whose transcription is induced during fruit and floral senescence. *J. Mol. Biol.* **222**:937-961.
16. Sato, T. and Theologis, A. (1989) Cloning the mRNA Encoding 1-aminocyclopropane-1-carboxylate synthase, the key enzyme for ethylene biosynthesis in plants. *Proc. Natl. Acad. Sci. USA* **86**:6621-6625.
17. Sheehy, R. E., Kramer, M. and Hiatt, W. R. (1988) Reduction of polygalacturonase activity in tomato fruit by antisense RNA. *Proc. Natl. Acad. Sci. USA* **85**:8805-8809.
18. Smith, C. J. S., Watson, C. F., Ray, J., Bird, C. R., Morriss, P. C., Schuch, W. and Grierson, D. (1988) Antisense RNA inhibition of polygalacturonase gene expression in transgenic tomatoes. *Nature* **334**:724-726.
19. Spanu, P., Reinhardt, D. and Boller, T. (1991) Analysis and cloning of the ethylene-forming enzyme from tomato by functional expression of its mRNA in *Xenopus laevis* oocytes. *EMBO J.* **10**:2007-2013.
20. Theologis, A. (1992) One rotten apple spoils the whole bushel: the role of ethylene in fruit ripening. *Cell* **70**:1-4.
21. Van Der Straeten, D., Van Wiemeersch, L., Goodman, H. M. and Van Montagu, M. (1990) Cloning and sequence of two different cDNAs encoding 1-aminocyclo-propane-1-carboxylate synthase in tomato. *Proc. Natl. Acad. Sci. USA* **87**:4859-4863.
22. Yang, S. F. and Hoffman, N. E. (1984) Ethylene biosynthesis and its regulation in higher plants. *Ann. Rev. Plant Physiol.* **35**:155-189.

GENETIC DISSECTION OF SIGNAL TRANSDUCTION PATHWAYS

THAT REGULATE *CAB* GENE EXPRESSION

Joanne Chory, Lothar Altschmied, Hector Cabrera,
Hsou-min Li, and Ronald Susek

Plant Biology Laboratory
The Salk Institute for Biological Studies
P.O. Box 85800
San Diego, CA 92186-5800

INTRODUCTION

In developing leaves, photosynthetically competent chloroplasts arise from small, undifferentiated proplastids that are present in meristematic cells. This process, called greening, involves the coordinate regulation of many nuclear and chloroplast genes (Mullet, 1988). The cues for the initiation of this developmental program are both extrinsic and intrinsic.

Of the various factors involved in chloroplast differentiation, the role of light is the best understood. Though several regulatory photoreceptors are involved in the perception of light, only the red-, far-red-light photoreceptors, called phytochromes, have been studied in detail for their role in leaf and chloroplast development. Light control of nuclear gene expression for chloroplast constituents appears to be mainly mediated by phytochrome (Silverthorne and Tobin, 1987). However, for maximal expression of light-regulated genes, the concerted action of phytochrome and a blue-light absorbing receptor appears to be required (Fluhr and Chua, 1986).

In addition to light, organ-, cell-, and developmental stage-specific signals have been implicated in chloroplast biogenesis (Mullet, 1988). Inhibition of leaf development leads to a simultaneous inhibition of chloroplast development and plastid transcription activity, suggesting that the development of the chloroplast is regulated in part by intrinsic signals that control leaf differentiation (Mullet, 1988). Further, only particular cell-types, the mesophyll cells, house chloroplasts, so presumably, cell-specific signals are also important determinants of chloroplast biogenesis. Finally, the developmental stage of the chloroplast itself is important for the expression of nuclear genes for chloroplast-destined proteins (e.g., Mayfield and Taylor, 1987).

These physiological data indicate that the regulation of greening is complicated. It involves the excitation of photoreceptors by light, transduction of this signal through other factors, and ultimately, increased expression of nuclear and chloroplast genes. Equally important are the developmental cues that control leaf morphology and cell-type-specific chloroplast biogenesis. The factors that mediate these responses are unknown.

To help elucidate the signal transduction pathways that lead to chloroplast biogenesis, we have undertaken a comprehensive genetic analysis of photomorphogenesis in the small cruciferous plant, *Arabidopsis thaliana*. Our initial efforts concentrated on morphological screens from which we and others identified photoreceptor and transduction mutants (for review, see Chory, 1991). However, the photomorphogenetic mutations are pleiotropic, affecting a large number of downstream light-regulated processes, including leaf and chloroplast development, pigment accumulation, and expression of light-regulated nuclear and chloroplast genes (Chory et al., 1989a, 1989b, 1991, Chory and Peto, 1990). Here we describe a second genetic approach that focuses on one particular downstream light-regulated response, the transcription of a chlorophyll a/b binding protein promoter, *cab3*. This molecular genetic approach allows us to identify mutants by aberrant gene expression patterns, rather than by predicted phenotype, and should be applicable to the study of other signal transduction pathways.

RESULTS

Strategy

We chose the nuclear *cab3* gene of *Arabidopsis* as an indicator of light-regulated, developmental gene expression (Leutwiler et al., 1986). The transcription of *cab* genes is tightly regulated. These genes are transcribed only in the light in green plants, with little or no detectable levels in etiolated seedlings. In addition, *cab* genes are expressed in a tissue-specific manner, transcripts being most abundant in leaves and lower or undetectable in other organs. Finally, the developmental stage of the chloroplast itself appears to regulate the expression of *cab* genes. For example, in photooxidative mutants of maize or in a variety of plants where chloroplast development is arrested with an inhibitor, *cab* genes are not transcribed (Mayfield and Taylor, 1987, Taylor, 1989). Thus, *cab* gene transcription is regulated by light, intrinsic developmental signals, and must also be sensitive to signals originating from the chloroplast itself.

The *cab* promoter-marker gene chimeras needed for these studies have been introduced into plants at a single site in each transgenic line. The transcriptional chimeras (Fig 1) contain a fully regulated *cab3* promoter sequence fused to either of two selectable marker genes: (a) the *hph* (hygromycin phosphotransferase) gene, which confers hygromycin resistance allowing for positive selection strategies (Gritz and Davies, 1983); or (b) the *Arabidopsis adh* (alcohol dehydrogenase) gene which allows either positive or negative selection (Chang and Meyerowitz, 1986). Both constructs show low basal expression from the *cab3* promoter and carry a screenable marker *uidA* gene (*E. coli* B-glucuronidase, GUS) under control of a second *cab* promoter (Jefferson, 1987). Our strategy is to mutagenize and select for plants which aberrantly express the marker transgenes from the *cab3* promoter under a variety of conditions. Following selection, we also screen for GUS activity, which is under the control of a second *cab3* promoter contained in the construct. This step is important so that true signal transduction mutants can be distinguished from *cis*-acting promoter mutations.

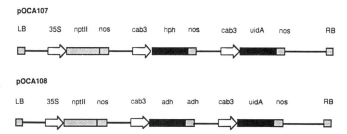

Figure 1. pOCA107 allows selection for hygromycin resistance in the dark with subsequent screening for GUS activity. *uidA* is the gene for the B-glucuronidase enzyme from *E. coli* (Jefferson, 1987). Many colorimetric and fluorometric substrates are available which make GUS screening feasible on small amounts of tissue. In pOCA108, either a positive or negative selection can be performed for the presence or absence of the alcohol dehydrogenase (ADH) gene. As for pOCA107, screening for GUS activity subsequent to selection should help delineate promoter mutations from *trans*-acting mutations. In both cases, the selection for the transformation event is by selection for kanamycin resistance, which is expressed from a highly and constitutively expressed plant promoter from cauliflower mosaic virus (35S).

We first transformed wild-type *Arabidopsis* with pOCA107 and an *adh* null mutant with pOCA108 constructions and obtained several independent transgenic lines. These lines were (a) diploid and (b) showed proper light/dark and tissue-specific expression of the *cab3* promoter. We focused our initial studies on pOCA107 lines. One of these lines, (pOCA107-2), was homozygous for the insertion and displayed tight light/dark, tissue-specific, and chloroplast-dependent regulation of the *cab3-uidA* transgene (Table 1).

Table 1. Expression of *cab3* promoter in pOCA107-2 line.

GUS Units[1]		Ratio
light-grown: 43,000	etiolated: 370	120
green leaves: 32,000	roots: 450	71
-norflurazon: 33,000	+norflurazon: 60	550

[1]GUS units are pmol 4-methylumbelliferone per min per mg of protein. Norflurazon is a herbicide that inhibits chloroplast development.

The pOCA107-2 transgene was mapped to chromosome 2, position 15.0 on the morphological marker map (Koornneef, 1990). We also mapped the transcription start site of the chimeric *cab-hph* fusion gene using primer extension to show that the mRNA was initiated at the proper start site and to show that it was properly regulated by light (Fig 2). 50,000 seeds were mutagenized with ethyl methanesulfonate (EMS), the plants were selfed, and the M2 seeds collected in pools of 200 families. We have been screening for *trans*-acting regulatory mutations that affect the expression of the two *cab* promoters in the pOCA107-2 line.

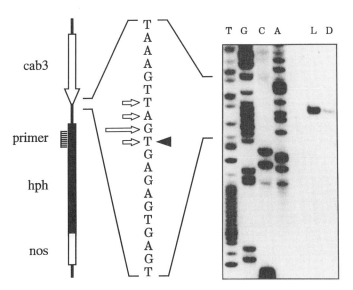

Figure 2. Light-dark regulation of the *cab3-hph* fusion in the pOCA107-2 transformant analyzed by primer extension analysis. The black arrow indicates the transcription start site, which is the same in this fusion construction as one that was published previously for the intact gene (indicated by the open arrows) (Karlin-Neumann et al., 1988).

We have looked for and obtained mutants where the *cab3* promoter is aberrantly transcribed in each of three situations. Specifically, we searched for mutants where the *cab* promoter was transcribed either: (1) in the dark; (2) in the wrong organs (e.g., roots); or (3) in the presence of inhibitors that eliminate chloroplast function. We designed our experiments such that all three classes of mutants would represent gain-of-expression of the *cab* promoter, which is important because the greening process is essential. The results from each screen are described separately.

Trans-acting Mutations that Affect the Etiolated Transcription Levels of *cab*3

One hundred thousand M2 seeds from 200 families were screened for hygromycin-resistance in the dark. 500 seeds from each family were germinated on a 90 mm Petri dish containing synthetic medium plus hygromycin (40 µg/ml) for 8 days in the dark. After 8 days, the plates were removed from the dark growth chamber, and were scored for growth. Seedlings from the unmutagenized pOCA107-2 parent had an average hypocotyl length of 1-4 mm in these conditions. 600 putative mutants were picked that had a hypocotyl longer than 6 mm. Of these 600, 92 survived and set seeds (M3). M3 seedlings from the putative mutants were assayed for GUS activity in the light and dark. An increased amount of GUS activity in the dark would distinguish a *trans*-acting mutation from a *cab*3 promoter mutation. Of the M3 seedlings examined, 7 also had a greater than 3-fold increase in GUS activity in the dark. All seven are from different M2 pools, indicating that they are independent mutational events.

As can be readily seen in the data presented above, only one in six of the putative mutants set seeds. One explanation for the poor survival rate was that the concentration of hygromycin used for the original selection was too high. We therefore re-screened 24 M2 families that contained putative mutants that did not survive the original screen on 20 µg/ml hygromycin. From this second screen, an additional five mutant lines were recovered that displayed a greater than 3-fold increase in GUS activity in the dark. We have designated the 12 new mutants as *doc* (*d*ark *o*verexpression of *c*ab).

We have backcrossed these mutant lines to either the parent pOCA107-2 line or wild-type Columbia and have shown that all *doc* mutants examined to date are single gene recessive mutations, with the exception of *doc*-473 which is a dominant mutation. We are currently performing complementation analysis to determine how many genes these mutations define. These mutants will be further characterized with respect to other aspects of light-regulated seedling development, as we have previously done with the photomorphogenetic mutants.

Trans-acting Mutations that Affect the Tissue-Specific Transcription of *cab*3

M2 seeds (200,000) were screened for root elongation after growth for 14 days in the light on synthetic medium containing 20 µg/ml hygromycin. Under these conditions, the pOCA107-2 control plants had short roots many of which did not contact the agar surface. Six hundred and seventeen putative mutants were picked that had long, branched root growth into the hygromycin-containing medium. A small piece of root from the putative mutants was excised and placed into a solution containing 5-Bromo-4-chloro-indolyl β-D-glucuronide (X-gluc), which allows for histochemical staining of GUS activity. Of the 617 M2 seedlings, 40 also showed X-Gluc staining in the roots, indicating a possible *trans*-acting mutation. Five of these gave rise to M3 seedlings with a hygromycin-resistant root phenotype and had weak X-gluc staining in the roots. However, the endogenous *cab* mRNAs did not accumulate to high levels in these mutants. To date, *det*1 alleles are the only mutants that show high inappropriate levels of *cab* mRNAs in root tissue (Chory and Peto, 1990). Since *det*1 alleles are simple to screen morphologically, we have abandoned the root hygromycin screen for the present time.

Mutations that Affect Signalling Between the Nucleus and Chloroplast

The herbicide norflurazon is known to block carotenoid accumulation, which results in photobleaching and the inhibition of chloroplast development in bright light. In a variety of plants that have been treated with norflurazon, the *cab* genes are not expressed. This has led to the hypothesis that a signal originating from the chloroplast is necessary for optimal transcription or mRNA accumulation for nuclear genes encoding proteins destined for the chloroplast (Taylor, 1989). Nothing is known about the nature of the signal, or the signal transduction pathway(s) involved in coupling gene expression in the chloroplast and nucleus during photosynthesis. We have used the mutagenized pOCA107-2 transgenic line to select for mutations in which the *cab*3 promoter is expressed at high levels in the presence of norflurazon. 100,000 M2 seeds, from 200 families, were grown for 10 days in bright light conditions in the presence of 5×10^{-6}M norflurazon and 20 µg/ml hygromycin. The visual screen is for photobleached plants that look larger and healthier (i.e., hygromycin-resistant) than the pOCA107-2 control seedlings. A cotyledon is "snipped" off each seedling and stained in X-gluc to test for expression of the second *cab* promoter. Of the original 664 plants selected, 227 had high levels of GUS activity, and 138 recovered and were fertile. M3 seeds from each individual M2 putative mutant were re-screened. Progeny from fourteen M2 plants are both hygromycin-resistant and GUS+, indicating possible *trans*-acting mutations. We have grown seedlings from the 14 putative mutant classes on norflurazon and isolated RNA. Northern blot analysis indicates that the endogenous *cab* genes are expressed in 11 of the 14 mutant lines in the presence of norflurazon. In contrast, we detect no *cab* RNA when wild-type plants are treated with norflurazon. Each of the 11 mutations represents an independent allele since each originated from a different M2 seed pool. To date, the 11 mutations define a minimum of 3 complementation groups and all the mutations are recessive when backcrossed to wild-type lines.

The RNA gel blots were reprobed with a second nuclear gene, *rbc*S, and two chloroplast genes, *psa*A-B and *rbc*L. For two of the eleven mutants, we detected no *rbc*L mRNA, and for the remaining 9, there was variable expression of *rbc*L. For *psa*A-B, there was less than 5% of wild-type activity in all 11 of the norflurazon-treated mutant lines. We have therefore designated these new mutants as *gun* (*g*enomes *un*coupled). For at least two of the *gun*1 alleles (*gun*1-1, *gun*1-2) our results indicate that the expression of *cab* genes has been totally uncoupled from the expression of both *rbc*L and *psa*A. We are focusing our current biochemical and genetic studies on these two uncoupled mutants.

Use of pOCA108 to Select for Mutations Involved in the Down Regulation of *cab*3 Transcription

We have recently established the conditions for screening for down-regulation of *cab*3 promoter expression using transgenic R002 lines that contain pOCA108. R002 is an *adh* null mutant isolated as an allyl alcohol resistant line in the mutagenized Bensheim background (Jacobs et al., 1988). In the pOCA108-1 line, ADH and GUS activities are expressed in a light-regulated and tissue-specific pattern that corresponds to expression of the *cab*3 promoter (data not shown). To perform screens for down-regulation of the *cab*3 promoter, seeds are germinated and grown in liquid for 5 days. At that time, the growth medium is exchanged for medium containing 1.5 mM allyl alcohol and the seedlings are treated for 2.5 h. After removal of allyl alcohol, the seedlings are returned to synthetic medium and allowed to grow for an additional day. Control pOCA108-1 seedlings are bleached and killed within 48 h of the allyl alcohol treatment. In contrast, R002 seedlings are completely resistant to the identical allyl alcohol treatments. The basis for selection is the fact that allyl alcohol, when added to the growth medium, is oxidized by alcohol dehydrogenase to acrolein, a toxic aldehyde that inactivates various proteins, binds to nucleic acids, and kills cells (Rando, 1974, Izard and Liberman, 1978). Fifty thousand seeds were recently mutagenized with either EMS or gamma rays and M2 families have been harvested. We are currently screening the families for *trans*-acting mutations that define genes involved in the positive regulation of *cab*3 gene expression.

CONCLUSIONS

In summary, we have isolated mutants that have aberrant *cab* promoter expression under any of several conditions, and have demonstrated for the first time in plants the feasibility of a strategy that should be applicable to a variety of regulated promoters. We are currently analyzing these new mutants, both genetically and biochemically. First, we will establish how many complementation groups define each class of mutant; then we will determine if we have identified some of the same genes using the different strategies. Finally, we will construct double mutants between these new mutants and the previously identified photomorphogenetic mutants to determine the number and sequence of events involved in early gene expression associated with light-regulated seedling development in *Arabidopsis*.

ACKNOWLEDGEMENTS

We thank Dr. Martin Yanofsky for the gift of the *adh* cloning cassette used in the pOCA108 construct. This work was supported by grants from the Department of Energy and the BRSG program to J.C. Initial experiments were performed by J.C. while she was a postdoctoral fellow in Dr. Frederick Ausubel's laboratory in the Department of Molecular Biology at Mass. General Hospital and were funded by a grant from Hoechst AG. Lothar Altschmied was a fellow of DFG and The Human Frontier Science Program, Hector Cabrera is a Fulbright fellow, Hsou-min Li is a fellow of the Samuel Roberts Noble Foundation, and Ronald Susek was partially supported by a grant from the Charles and Ruth Billingsley Foundation.

REFERENCES

Chang, C., and Meyerowitz, E.M., 1986, Molecular cloning and DNA sequence of the *Arabidopsis thaliana* alcohol dehydrogenase gene, *Proc. Natl. Acad. Sci.* 83:1408.

Chory, J., 1991, Light signals in leaf and chloroplast development: photoreceptors and downstream responses in search of a transduction pathway, *New Biol.* 3:538.

Chory, J., Nagpal, P., and Peto, C., 1991, Phenotypic and genetic analysis of *det2*, a new mutant that affects light-regulated seedling development in *Arabidopsis*, *Plant Cell.* 3:445.

Chory, J. and Peto, C., 1990, Mutations in the *DET1* gene affect cell-type-specific expression of light-regulated genes and chloroplast development in *Arabidopsis*, *Proc. Natl. Acad. Sci.* 87:8776.

Chory, J., Peto, C., Ashbaugh, M., Saganich, R., Pratt, L., and Ausubel, F., 1989a, Different roles for phytochrome in etiolated and green plants deduced from characterization of *Arabidopsis thaliana* mutants, *Plant Cell.* 1:867.

Chory, J., Peto, C., Feinbaum, R., Pratt, L., and Ausubel, F., 1989b, *Arabidopsis thaliana* mutant that develops as a light-grown plant in the absence of light, *Cell.* 58:991.

Colbert, J., 1985, Molecular biology of phytochrome, *Plant Cell Environ.* 11:305.

Fluhr, R., Chua, N.H., 1986, Developmental regulation of two genesencoding RubisCO small subunit in pea and transgenic tobacco plants: phytochrome response and blue-light induction, *Proc. Natl. Acad. Sci.* 83:2358.

Gritz, L., and Davies, J., 1983, Plasmid-encoded hygromycin B resistance: the sequence of hygromycin B phosphotransferase gene and its expression in *Escherichia coli* and *Saccharomyces cerevisiae*, *Gene* 25:179.

Izard, C., and Liberman, C., 1978, Acrolein, *Mutat. Res.* 47:115.

Jacobs, M., Dolferus, R., and van den Bossche, D., 1988, Isolation and biochemical analysis of ethyl methanesulfonate-induced alcohol dehydrogenase null mutants of *Arabidopsis thaliana* (L.)Heynh., *Biochem. Genet.* 26:105.

Jefferson, R.A., 1987, Assaying chimeric genes in plants: the GUS gene fusion system, *Plant Mol. Biol. Reporter.* 5:387.

Karlin-Neumann, G.A., Sun, L., and Tobin, E., 1988, Expression of light-harvesting chlorophyll a/b protein genes is phytochrome-regulated in etiolated *Arabidopsis thaliana* seedlings, *Plant Physiol.* 88:1323.

Koornneef, M., 1990, Linkage map of *Arabidopsis thaliana*, in: "Genetic Maps: Locus Maps of Complex Genomes, " 5th ed., S.J. O'Brien, ed., Cold Spring Harbor Press, Cold Spring Harbor, New York.

Leutwiler, L., Meyerowitz, E.M., and Tobin, E.M., 1986, Structure and expression of three light-harvesting chlorophyll a/b binding protein genes in *Arabidopsis thaliana*, *Nucl. Acids Res.* 14:4051.

Mayfield, S.P., and Taylor, W., 1987, Chloroplast photooxidation inhibits the expression of a set of nuclear genes, *Mol. Gen. Genet.* 208:309.

Mullet, J.E., 1988, Chloroplast development and gene expression, *Annu. Rev. Plant Physiol. Plant Mol. Biol.* 39:475.

Rando, R.R., 1974, Allyl alcohol induced irreversible inhibition of yeast alcohol dehydrogenase, *Biochem. Pharmacol.* 23:2328.

Silverthorne, J., and Tobin, E., 1987, Phytochrome regulation of nuclear gene expression, *Bioessays.* 7:18.

Taylor, W.C., 1989, Regulatory interactions between nuclear and plastid genomes, *Annu. Rev. Plant Physiol. Plant Mol. Biol.* 40:211.

THE ROLE OF *KN1* IN PLANT DEVELOPMENT

Sarah Hake

USDA/U.C. Berkeley Plant Gene Expression Center
800 Buchanan Street
Albany, CA 94706

Cell communication is an ubiquitous feature of plant development; plant cells divide, expand, and differentiate in concerted action. Genetically distinct cells adopt alternate fates depending on their neighbors, in fact, the fates of all cells are most decisively determined by their neighbors. The question remains, not whether cell communication signals play a role, but what are these signals and how do they function? I will discuss areas of development that high-light the role of cell communication, and then suggest that *KN1* may be one of the mediators.

INTRODUCTION

Plants proceed through a series of developmental steps that begin with formation of the meristem. We do not know what cells are recruited to become meristem. Clonal analysis in maize suggests that the meristem becomes determined at 8-10 days post-pollination from 100-200 cells of the embryo (Poethig et al., 1986). Since the meristem cells can be related to cells that give rise to the scutellum, or nutritive tissue (Poethig et al., 1986), it suggests that pre-meristem cells are not special cells. Meristem cells, most likely, happen to be in a certain place at the right time. The meristem becomes layered during its embryonic development. The outer layer of the meristem, or L1, contributes primarily to the epidermis, the inner layers, L2 and L3, contribute to the bulk of the plant body. However, these layers are not fixed, as cells from the L2 divide into the L1 and vice versa. Having changed layers, the cells quickly take on the fates of their new position.

Leaves arise as lateral organs from the flanks of the meristem. In a similar fashion, floral organs are initiated from the flanks of floral meristems. In a manner of a few divisions or a slight twist of position, a cell's fate is dramatically altered from indeterminate to determinate. It is no longer capable of totipotent growth, but rather is capable of differentiating complete new phenotypes. What possible mechanisms may guide the recruitment and evolution of cells into lateral organs? One possibility proposed by Green and his coworkers is that physical stresses on the apex cause bucklings, the bucklings then lead to the next primordia (Green, 1989; Jesuthasan and Green, 1989). Such a mechanism attributes a passive role to the cell, and suggests that gene expression of that particular cell would occur after the physical components have initiated the process. Another possibility is that initiation sites are induced by some sort of chemical signals. The signal is likely to emanate from the meristem itself. There must be a genetic component to the signals, since different species have different patterns of initiation, but the signal or receiving mechanisms can also be physically disturbed by surgical interference (see references in Hake and Sinha, 1991).

Organ formation initially involves supracellular mechanisms of regulation. The groups of cells that will form an organ must act in concert, although there might be different outcomes if the cells were not placed in one organ together. For example, the dwarf mutation *D8* conditions a severely dwarfed phenotype, is unresponsive to gibberellins and produces anthers in the normally all female ear. Using a linked albino marker, Harberd and Freeling (1989) examined sectors that had lost the dominant *D8* locus following irradiation-induced chromosome breakage. White sectors in the cob, which would carry the wildtype *d8* genotype, were not anther-eared, suggesting this phenotype is tissue autonomous. Also, sectors in the tassel produced white tassel branches that were longer than the other dwarfed branches. However, 29 of 32 sectors in the leaf did not affect the shape of the leaf. In one larger sector, there was buckling along the *D8* and wildtype border. It appears that if the sector includes a certain percentage of the cells, it can autonomously regulate its final size, otherwise its behavior is dictated by its neighbor cells.

Supracellular control continues during the early stages of histogenesis. The production of the ligule is an example of cell communication. The ligule forms at the junction of leaf blade and leaf sheath, at an early stage of leaf development. Cells forming the ligule are defined by their position on the leaf at an early stage (Hake et al., 1985; Sylvester et al., 1990). Genetic mutations can interrupt the process of defining the position of the ligule, suggesting that the genes they define act either to transduce the signal or accept it (Becraft et al., 1990). Vein differentiation is another example of a relatively late event in which cell neighbors play a role in fate determination. Cells become mesophyll or bundle sheath cells depending on their position next to the vein, and not depending on their cell lineage (Langdale et al., 1989). The ability of a cell to influence its neighbor becomes more and more restricted as differentiation continues.

KN1 EXPRESSION PATTERNS IN WILDTYPE

The *KN1* cDNA is predicted to encode a polypeptide that contains a homeodomain (Vollbrecht et al., 1991). The homeodomain is a DNA-binding motif that had been well characterized in animal systems and is often encoded by genes that are critical in development (for review, see Akam, 1987; Ingham, 1988). Immunolocalization of an antibody to *KN1* demonstrates that *KN1* expression patterns are normally restricted to the meristem and the shoot apex. The protein is downregulated in cells that will initiate leaves, as well as developing leaves (Figure 1) (Smith et al., 1992). The downregulation of *KN1* occurs even prior to the time that a leaf primordium is visible as separate from the meristem, and is in fact, the earliest event known in leaf initiation. What causes the loss of *KN1*? How is *KN1* regulated? The answers to these questions will bring us closer to the regulation of leaf initiation. However, the detection of an early cellular marker for leaf initiation does suggest that physical constraints may not be the initiating event.

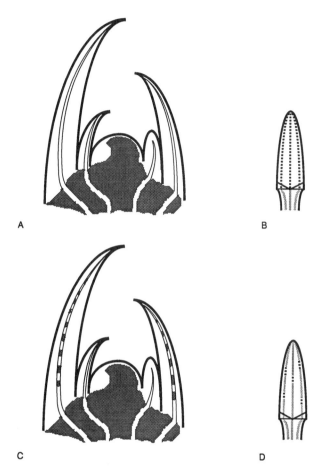

Figure 1. Diagram that illustrates the *KN1* immunolocalization patterns in young maize plants. (A) Longitudinal sections of wild-type seedlings reveal that the meristem and shoot are stained with *KN1* antibody, indicated by dots, but the leaf primordia and the lateral veins within the shoot are not stained. (B) Wild-type leaves are characterized by distinct blade (squares) and sheath (fine dots) components. (C) In *KN1* mutants, ectopic staining is seen in lateral veins of leaf primordia in addition to the meristems and shoot staining. (D) Lateral veins of the leaf blade that express *KN1* differentiate into ligule or sheath-like cells, indicated by the presence of the sheath pattern in black.

Figure 2. Leaf blade of a *Kn1-N* mutant plant. The knots are focused on the lateral veins.

Expression in the shoot apex is strongest in the region close to the meristem, prior to expansion. Most of the cells express *KN1* except certain vein cells. Lateral veins are initiated in the shoot and differentiate into the leaf as the leaf grows. They are not stained in the leaf or shoot. Veins that are initiated in the leaf and differentiate basipetally are called intermediate veins. They are stained in the shoot, but not in the leaf. A possible explanation for why one set of veins is stained and not the other may relate to the timing of their initiation. Lateral veins are initiated early, as the leaf itself initiates. Perhaps the lateral veins play a role in the initiation itself, and thus are downregulated in concert with the leaf cells. Another possibility is suggested by the finding that the staining in the intermediate veins dissapears as cells differentiate, suggesting that the appearance of *KN1* is correlated with an undetermined state. It may be that lateral veins in the shoot are much more determined than intermediate veins and, thus, never express *KN1*.

DOMINANT *Kn1* MUTATIONS

Kn1 was initially defined by dominant mutations that affect the development of the leaf (Bryan and Sass, 1941; Gelinas et al., 1969; Freeling and Hake, 1985). The mutations result from insertions or rearrangements to non-coding regions of the gene (see Hake, 1992 for review) and do not affect the RNA and protein product. The maize mutations all result in similar phenotypes that specifically affect the leaf blade. Knots are focused on lateral veins and occur somewhat sporadically (Figure 2). They result when a group of cells continues to grow even though it is surrounded by non-growing cells. In addition to knots, lateral veins take on sheath-like characteristics (N. Sinha, unpublished data). The ligule of knotted leaves is often displaced into the blade, found ectopically in the blade, or missing altogether. The many phenotypic abnormalities have a common theme. The cells of the leaf blade have taken on fates of cells found in lower regions of the leaf.

The *KN1* protein is ectopically expressed in developing lateral veins of mutant plants. It is not immunologically detectable in wildtype leaves. The expression is visible fairly late in development of the leaf. The expression begins sporadically, but eventually all lateral veins are affected. Somewhat later, the intermediate veins also express *KN1* (Smith et al., 1992). The early expression of *KN1* in lateral veins of leaf blades correlates well with the mutant phenotype. The knots, ligule displacement and sheath characteristics are primarily restricted to the lateral veins. The intermediate veins are mildly affected, never showing knots or ligule, but expressing a mild sheath expression pattern (N. Sinha, unpublished data). The *KN1* expression in the intermediate veins may fail to cause a response because these cells have ceased dividing and have begun to differentiate compared to their lateral vein counterparts.

The protein expression patterns also substantiate earlier clonal analysis. Sectors that carried the *Kn1* mutation in the epidermis had no mutant phenotype (Hake and Freeling, 1986), whereas sectors with *Kn1* only in the inner layer were mutant (Sinha and Hake, 1990). We now realize that the epidermal layer probably had no effect because it does not express the protein. Nevertheless, the conclusion that the inner layer affects the outer layers is still significant. The expression of *KN1* in the inner vein cells causes all surrounding cells to differentiate in new ways. Most striking is the differentiation of epidermal cells into ligule. It is not likely that *Kn1* directly acts on ligule-forming genes since *Kn1* is not expressed in the ligule-forming cells. It is also absent from cells that normally differentiate into ligule. It is more likely that *Kn1* misregulates more global regulators that in turn regulate specific genes such as defined by the *liguleless* mutations.

ROLE OF *KN1*

Although we may have a conceptual framework with which to understand the dominant mutant phenotype, we still have no answers for what role *KN1* plays in its wildtype context. What is the loss of function phenotype? It is possible that the absence of *KN1* has no effect, that another gene covers for *KN1* and there is no phenotype. While this possiblity is kept in mind, we are pursuing experiments to test whether *KN1* is a required gene with a lethal null phenotype.

REFERENCES

Akam, M., 1987, The molecular basis for metameric pattern in the *Drosophila* embryo, *Development* 101:1-22.

Becraft, P. W., Bongard-Pierce, D. K., Sylvester, A. W., Poethig, R. S., and Freeling, M., 1990, The *liguleless-1* gene acts tissue specifically in maize leaf development, *Dev. Biol.* 141:220-232.

Bryan, A. A., and Sass, J. E., 1941, Heritable characters in maize, *J. Hered.* 32:343-346.

Freeling, M., and Hake, S., 1985, Developmental genetics of mutants that specify *Knotted* leaves in maize, *Genetics* 111:617-634.

Gelinas, D., Postlethwait, S. N., and Nelson, O. E., 1969, Characterization of development in maize through the use of mutants. II. The abnormal growth conditioned by the *Knotted* mutant, *Am. J. Bot.* 56:671-678.

Green, P. B., 1989, Shoot morphogenesis, vegetative through floral, from a biophysical perspective, *in*: "Plant Reproduction: From Floral Induction to Pollination," E. Lord, and G. Bernier, eds., pp. 58-75, Amer. Soc. Plant Physiol., Rockville, MD.

Hake, S., 1992, Unraveling the knots in plant development, *Trends Genet.* 8:109-114.

Hake, S., Bird, R. M., Neuffer, M. G., and Freeling, M., 1985, Development of the maize ligule and mutants that affect it, *in*: "Plant Genetics," M. Freeling, ed., pp. 61-71, Alan R. Liss, Inc, New York.

Hake, S., and Freeling, M., 1986, Analysis of genetic mosaics shows that the extra epidermal cell divisions in *Knotted* mutant maize plants are induced by adjacent mesophyll cells, *Nature* 320:621-623.

Hake, S., and Sinha, N., 1991, Genetic analysis of leaf development, *Ox. Surv. Plant Mol. Cell. Biol.* 7:in press.

Harberd, N. P., and Freeling, M., 1989, Genetics of dominant gibberellin-insensitive dwarfism in maize, *Genetics* 121:827-838.

Ingham, P. W., 1988, The molecular genetics of embryonic pattern formation in *Drosophila*, *Nature* 335:25-34.

Jesuthasan, S., and Green, P. B., 1989, On the mechanism of decussate phyllotaxis: biophysical studies on the tunica layer of *Vinca major*, *Amer. J. Bot.* 76:1152-1166.

Langdale, J. A., Lane, B., Freeling, M., and Nelson, T., 1989, Cell lineage analysis of maize bundle sheath and mesophyll cells, *Dev. Biol.* 133:128-139.

Poethig, R. S., Coe Jr., E. H., and Johri, M. M., 1986, Cell lineage patterns in maize embryogenesis: a clonal analysis, *Dev. Biol.* 117:392-404.

Sinha, N., and Hake, S., 1990, Mutant characters of *Knotted* maize leaves are determined in the innermost tissue layers, *Dev. Biol.* 141:203-210.

Smith, L., Greene, B., Veit, B. and Hake, S., 1992, A dominant mutation in the maize homeobox gene, *Knotted-1*, causes its ectopic expression in leaf cells with altered fates, *Development* in press.

Sylvester, A. W., Cande, W. Z., and Freeling, M., 1990, Division and differentiation durng normal and *liguleless-1* maize leaf development, *Development* 110:985-1000.

Vollbrecht, E., Veit, B., Sinha, N., and Hake, S., 1991, The developmental gene *Knotted-1* is a member of a maize homeobox gene family, *Nature* 350:241-243.

GENETIC ANALYSIS OF MERISTEM STRUCTURE AND FUNCTION IN *ARABIDOPSIS THALIANA*

M. Kathryn Barton

Department of Biology
University of Pennsylvania
Philadelphia, PA 19104

INTRODUCTION

Meristems are small regions of undifferentiated, dividing cells located at the tips of roots and shoots. Shoot and root apical meristems are formed in the embryo and upon germination give rise to all the above and below ground structures of the plant respectively. Not only do these apical meristems themselves generate plant organs such as leaves, flowers, stems and roots but they typically also generate additional, or lateral, meristems.

Angiosperm shoot meristems can be described as consisting of two regions, a tunica and a corpus (Schmidt, 1924). The tunica consists of one or more layers of cells that cloak the summit of the meristem. The generation and the maintenance of these cell layers is dependent on the persistence of anticlinal divisions in this layer. (Anticlinal divisions are those in which the spindle is oriented parallel to the nearest surface. Periclinal divisions, mentioned below, are those in which the spindle is oriented perpendicular to the nearest surface.) Underlying the tunica is the corpus. The cells in the corpus do not exhibit any obvious stratification reflecting a greater freedom in choice of cell division plane in this region.

Although the significance of the tunica-corpus organization to meristem function is not known, this organization is reflected in the ultimate fates of the descendants of the meristem cells (e.g. Dulieu, 1968; Stewart and Burk, 1970). For instance, cells in the outer tunica layer typically give rise to the epidermis of the leaf. Likewise, cells of the inner tunica layers and the corpus typically give rise to progressively more internal tissues of the leaf. Although such generalizations can be made, it is important to note that it is not a cell's history that is important but rather its position. This is shown by the fact that if a cell from the outer tunica layer "invades" an inner layer, its progeny will take on fates characteristic of cells in the inner layer.

Cells in the meristem, or at least a subset thereof, exhibit the characteristics of what have been called stem cells in animal biology. These cells are defined as such by their ability to: proliferate, give rise to a variety of cell types, replenish themselves and regenerate the meristem upon injury (Potten and Loeffler, 1990). To avoid confusion, however, the term stem cell has been avoided in plant biology and replaced with the term apical initial.

Using plants mosaic for color markers, Stewart and Dermen (1975) surveyed a number of species and found that for each layer of the shoot meristem, there are from one to three apical initial cells. Consistent with results found in animal systems (Winton and Ponder, 1990), they found that the apical initials need not themselves show a rapid rate of division. In fact, they calculated that for California privet these cells divided only once every twelve days.

The fact that meristems are able to regenerate themselves when surgically manipulated in a variety of ways (reviewed by Sussex, 1989) argues strongly that there is extensive intercellular communication within the meristem. With the goal of ultimately understanding how the plant meristem works on a molecular level, I have undertaken a genetic analysis of meristem function using the small crucifer *Arabidopsis thaliana* as an experimental organism. By isolating mutants that either lack a meristem or exhibit altered meristem development I have sought to find genes whose products are important in the normal functioning and development of the meristem.

Several types of functions that are susceptible to mutation can be envisioned. These would include the initiation of the meristems, the maintenance of a proliferating population of cells once formed and the maintenance of the tunica-corpus organization within the meristem. In addition, one might expect to find genes involved in determining the placement and number of organs produced by the meristem as well as in specifying their identity.

I shall concentrate here on mutations at two loci, *shootless* and *rootless*, that affect meristem initiation in *Arabidopsis* and will briefly describe two other loci, *pinhead* and *altered phyllotaxis*, mutations in which affect the maintenance of the meristematic region and the placement of organs on the meristem respectively.

RESULTS AND DISCUSSION

Development of the Wild-type Shoot Apical Meristem

The vegetative meristem of *Arabidopsis thaliana* consists of two tunica layers overlying a rather shallow corpus (Vaughan, 1954; Miksche and Brown, 1965). The initiation of the shoot apical meristem in the embryo has been described (Barton and Poethig, in preparation). The pertinent events can be summarized as follows. Prior to the torpedo stage, three layers of cells at the presumptive shoot apex are formed. At or after the torpedo stages cells in these layers assume the division patterns characteristic of the meristem layers that they will give rise to. That is, cells in the epidermal layer and the subjacent layer divide anticlinally while cells in the lowest layer show a mixture of cell division planes - anticlinal, periclinal and oblique. Based on their positions in the embryo and on the inability of plant cells to migrate, it can be deduced that these correspond to the two layers of the tunica and the corpus respectively.

Mutants that Lack a Shoot Apical Meristem

Seedlings homozygous for the *shootless* mutation lack a shoot apical meristem. Sections of mutant embryos show that *shootless* embryos lack a shoot apical meristem (Barton and Poethig, in preparation) indicating that the defect in *shootless* mutants is a failure to initiate a shoot apical meristem. The configuration of cells at the apex of shootless mutants (three layers of cells deep and two to four cells in breadth) is the same as that found in wild-type torpedo stage embryos suggesting that the *shootless* gene product plays a role at the transition point in meristem development at which cells of these layers assume the division patterns that will characterize them later in life. It may in fact be that this transition in fact defines the initiation of the shoot apical meristem in the embryo. This is supported by the observation that lateral meristems in other species can also develop via a three layer structure. For instance, in quack grass, three layers of cells in the axils of leaves divide analogously to that described above for the shoot apical meristem of wild-type *Arabidopsis* (Sharman,1945). It remains to be seen whether lateral meristems in Arabidopsis develop through such a three layered intermediate and whether the *shootless* mutation affects the development of lateral meristems.

Whereas wild-type root explants will give rise to shoots in culture, root explants made from homozygous *shootless* tissue are not able to do so (Barton and Poethig, in preparation), arguing for a role of the *shootless* gene product in the formation of shoot apical meristems postembryonically as well as in the embryo.

Homozygous *shootless* mutant seedlings show an additional phenotype. A few days after germination the hypocotyls of *shootless* seedlings swell and at about 14 days, leaves burst from the swollen region. These leaves are somewhat abnormal in their shape and a normal shoot is not formed. This result suggests that the apical meristem in wild-type *Arabidopsis* inhibits the cells below it form dividing and redifferentiating.

Development of the Root Apical Meristem in Wild-type *Arabidopsis thaliana*

The root apical meristem is initiated during embryogenesis. The initiation of the root in the Arabidopsis embryo proceeds through a series of stereotyped and in some cases invariant divisions (Barton and Poethig, in preparation) and is similar to the development of the root primordium in other species of Brassica (e.g. Raghavan, 1990). The hypophysis, a cell derived from the suspensor, gives rise to the central portion of the root cap and to the very tip of the main body of the root itself. It does so through a series of transverse and longitudinal divisions. The lateral portion of the root cap arises from epidermal cells of the embryo. The generation of lateral root cap requires a switch from anticlinal to periclinal divisions in these epidermal cells.

Mutants that Lack a Root Apical Meristem

Seedlings homozygous for any one of three mutant alleles of the *rootless* locus lack root apical meristems (Barton and Poethig, in preparation). In addition, all three mutations are pleiotropic. Additional defects seen in a proportion of homozygous *rootless* seedlings are: one cotyledon instead of two; a short inflorescence that tapers to a spike; absence of a shoot apical meristem; a reduced hypocotyl.

When median longitudinal sections of *rootless* individuals are compared to those of wild-type, they show an altered organization of cells in the root primordium. The hypophysis instead of dividing both longitudinally and transversely divides only transversely. In addition the cells in the embryonic epidermis fail to switch from anticlinal to periclinal divisions to generate lateral root cap. This result emphasizes the importance of the control of cell division plane in plant architecture. Whether alterations in the division pattern of the hypophysis cause a defect in root formation, e.g. by failing to cause the proper localization of "root determinants" or whether these alterations are the effect of the failure to elicit a root program of differentiation in *rootless* mutants is not known.

Mutants that Fail to Maintain a Shoot Apical Meristem

Seedlings homozygous for the *pinhead* mutation are not able to maintain a shoot apical meristem. Instead, either a slender pin shaped structure roughly six cells in width, or a leaf is formed. The leaf may be normally shaped or it may be radially symmetrical and trumpet shaped. Thus, the *pinhead* locus is a candidate for a gene required to maintain a population of dividing, undifferentiated cells in the meristem. In the putative absence of the *pinhead* gene product, cells would immediately differentiate. The difference in whether a pin-like structure or a leaf develops could reflect the number of cells present in the meristem before it "runs out", more cells being present in the case where a leaf is formed.

Mutants with altered phyllotaxis

Seedlings homozygous for the *altered phyllotaxis (phl)* mutation typically have three cotyledons and subsequently produce leaves that are more tightly spaced than in wild-type. Snow and Snow (1933) performed experiments that suggested that the placement of new leaves was dependent on zones of inhibition surrounding preexisting leaf primordia. According to such a model, the zones of inhibition would be reduced in *phl* mutants, suggesting a decrease in an inhibitory signal in this mutant. In fact, in the most extreme *phl* phenotype, a single bowl-shaped cotyledon is formed. In addition, the phenotype of the *phl* mutants suggests that leaves and cotyledons share some of the same spacing signals. The adult *phl* plant shows a decrease in apical dominance, again suggesting a defect in intercellular communication.

CONCLUSION

Meristem development has been studied for many years yet little progress has been made towards an understanding of the molecular mechanisms underlying meristem structure and function (Sussex, 1989). Mutants with defects that are specific to meristem functioning will aid in the identification of genes critical to meristem development. These will provide powerful tools for ultimately understanding the structure and function of the plant meristem in molecular detail. It is likely that some of these genes will identify the signaling molecules and receptors responsible for the highly organized behavior of cells within the plant meristem.

REFERENCES

Dulieu, H. 1968 Emploi des chimeres chlorophyllienes pour l'etude de l'ontogenie foliaire. Bull.Soc.Bourgogne 25:1-60.

Miksche, J.P. and J.A.M. Brown, 1965, Development of vegetative and floral meristems of *Arabidopsis thaliana.* Amer.J.Bot. 533-537.

Potten, C.S. and M. Loeffler, 1990 Stem cells: attributes, cycles, spirals, pitfalls and uncertainties. Lessons for and from the crypt. Development 110:1001-1020.

Raghavan, V. 1990, Origin of the quiescent center in the root of *Capsella bursa-pastoris* (L.) Medik Planta 181:62-70.

Schmidt, A. , 1924 Histologische Studien an phanerogamen Vegetationspunkten. Bot.Arch. 8:345-404.

Sharman, B.C. 1945 Leaf and bud initiation in the Gramineae. Bot.Gaz.31:269-289.

Snow, M. and R. Snow, 1933, Experiments on phyllotaxis.II.The effect of displacing a primordium. Phil.Trans.R.Soc.London Ser.B 222:353-400.

Stewart, R.N. and L.G. Burk, 1970 Independence of tissues derived from apical layers in ontogeny of the tobacco leaf and ovary. Am.J.Bot.57:1010-1016.

Stewart, R.N. and H. Dermen, 1970 Determination of number and mitotic activity of shoot apical initial cells by analysis of mericlinal chimeras. Amer.J.Bot. 57:816-826.

Sussex, I.M., 1989, Developmental programming of the shoot meristem. Cell 56:225-229.

Vaughan, J.G. 1954, The morphology and growth of the vegetative and reproductive apices of *Arabidopsis thaliana* (l.) Heynh., *Capsella bursa-pastoris* (L.) Medic. and *Anagallis arvensis* L. Journ.Linn.Soc.Bot. 55:279-300.

Winton, D.J. and B.A.J. Ponder, 1990, Stem-cell organization in mouse small intestine. Proc.R.Soc.Lond.B 241:13-18.

AXILLARY BUD DEVELOPMENT IN PEA: APICAL DOMINANCE, GROWTH CYCLES, HORMONAL REGULATION AND PLANT ARCHITECTURE

Joel P. Stafstrom

Plant Molecular Biology Center and
Department of Biological Sciences
Northern Illinois University
DeKalb, IL 60115

APICAL DOMINANCE AND THE DEVELOPMENT OF APICAL MERISTEMS

Apical meristems of the shoot and root are responsible for building the vegetative plant body. Precise patterns of cell division and differentiation at the shoot apex result in the iterative formation of modules or phytomers comprised of a leaf, an axillary bud, a node and an internode (Sussex, 1989). Axillary bud meristems have the same potential for growth and development as the terminal meristem; however, most buds remain dormant and never realize this potential. As any gardener or keeper of house plants knows, removing the terminal bud promotes the growth of dormant axillary buds and gives rise to bushier plants. Control of axillary bud growth by the terminal bud is called apical dominance.

Axillary buds of the garden pea, *Pisum sativum* cv. Alaska, remain dormant indefinitely on intact plants but begin to grow within hours of decapitating the terminal bud. Many investigators have taken advantage of this strict developmental regulation to test hypotheses regarding the physiology and anatomy of bud growth and apical dominance (e.g., Phillips, 1975). Our primary goal has been to understand the biochemical and molecular events that occur within buds at each stage of their development. We have identified several growth-specific genes and have used these as molecular markers during cycles of bud growth and dormancy. We have found that dormant buds synthesize unique sets of proteins and mRNAs and that they are metabolically active. Whether a bud begins to grow and continues growing depends on information that it exchanges with the plant. While auxin and cytokinins appear to be involved in this long-range communication, it is clear that other components also are necessary for signal transduction within and between cells. The ultimate expression of bud growth is the formation of a new branch or shoot, which will in turn alter the plant's architectural form. In this paper, we review published and previously unpublished experiments and speculate on how differential bud development might influence plant form.

CYCLES OF GROWTH AND DORMANCY

The term "dormancy" has been applied to many distinct events during the plant life cycle. Different classes of dormancy have been defined by the organs affected, the conditions that promote dormancy or the sources of signals that promote dormancy. We refer to non-growing buds on intact plants as "dormant" but, as yet, we do not have a precise molecular description of this state nor do we know how it might be related to other classes of dormancy. Our working definition of dormancy is "the temporary suspension of visible growth of a plant structure containing a meristem" (Lang, 1987).

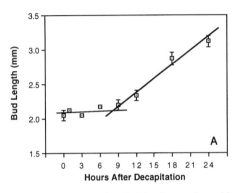 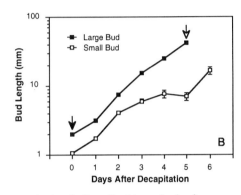

Figure 1. Kinetics of node-2 bud growth on Alaska pea plants. A. Growth of large buds after decapitation. There is a lag of about 8 h before growth begins. Mean +/- SE. B. Large and small buds begin to grow following decapitation (closed arrow). Growth of the small buds is reduced at 3 d and ceases after 5 d. Removing the large buds after 5 d (open arrow) allows small buds to resume growing. Mean +/- SE. From Stafstrom and Sussex, 1992. Copyright American Society of Plant Physiologists, used with permission.

Alaska pea seedlings contain four buds at the second node, all of which begin to grow when plants are decapitated. The largest bud nearly doubles in length and triples in mass 24 hours after decapitation. There is a lag of eight hours before growth can be detected (Figure 1A). The second largest bud, which we call the "small bud," grows rapidly for two days following decapitation and then slows down and ceases growing after four days (Figure 1B). Removing the large bud after 5 days allows the small buds to resume growing. Therefore, buds can cycle rapidly and repeatedly between states of growth and dormancy. It is known that rapid growth of a bud can inhibit buds at higher and lower nodes in pea (Husain and Linck, 1966) and at opposite nodes of *Alternanthera*, a decussate plant (Cutter, 1968). However, a detailed analysis of the transition from growth to dormancy has not been documented previously. In contrast to the cyclical pattern of vegetative meristem development, the development of vegetative meristems into a reproductive meristems is generally a unidirectional process (Bernier, 1988).

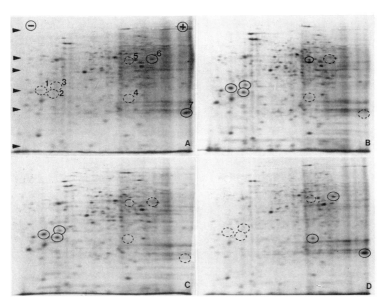

Figure 2. 2D-PAGE of *in vitro* translation products from axillary buds. A. D buds, intact plants. B. D->G buds, 6 hr after decapitation. C. G buds, 24 hr after decapitation. D. G->D small buds, 4 days after decapitation. The translation products mentioned in the text are numbered (in panel A) and circled with a solid line when present and a broken line when absent or greatly reduced. Molecular weight standards were 18.4, 29, 36, 43, 55 and 95.5 kDa (arrowheads in A). The isoelectric focusing range was between pH4.7 (-) and pH7.3 (+).

We have defined four stages of bud development. The dormant (D) and growing (G) states occur on intact plants and plants decapitated 24 hours earlier, respectively. Since buds can persist in these states for prolonged periods of time, they are considered to be stable. In addition, two unstable or transition states have been identified. The dormant-to-growing (D->G) transition occurs on decapitated plants before visible growth can be detected at eight hours. As we demonstrate below, growth-specific events can be detected much sooner. The growing-to-dormant (G->D) transition occurs in small buds four days after the main shoot is removed (see Figure 1B). Is there a physiological or metabolic rationale to indicate that these stages are distinct? Two-dimensional gel analysis of total and *in vivo* labeled proteins from D, D->G and G buds demonstrated that each synthesizes a unique set of proteins (Stafstrom and Sussex, 1988). Within three hours of decapitation, the protein profile of D->G transition buds was more similar to that of growing buds than that of dormant buds. Overall, the expression of approximately 10% of the proteins detected by each technique was altered qualitatively or quantitatively during the dormancy to growth transition. Dormant buds were found to incorporate labeled amino acids into protein at rates equivalent to growing buds. This rather surprising finding suggests that dormancy is an active process. Maintainance of dormancy may require the expression of dormancy-specific genes, the repression of growth-specific genes, or both.

All four bud stages were analyzed by *in vitro* translation and 2D gel electrophoresis to determine how steady-state mRNA levels change during a complete growth/dormancy cycle (Figure 2). A few translations products are highlighted to illustrate various patterns of expression. In general, D->G and G buds contained similar sets of translation products (spots 1, 2 and 3) as did D and G->D buds (spots 6 and 7). Thus, when buds re-enter the dormant state, growth-specific mRNAs decline in abundance and dormancy-specific mRNAs increase. Each transition stage also contained translation products that were not found in any other stage (spot 5 in D->G and spot 4 in G->D). Therefore, the molecular events that occur during exit from dormancy differ from those that occur during entry into dormancy. These results indicate that a single growth/dormancy cycle contains at least four distinct steps.

Thimann and Skoog (1934) found that adding auxin in lanolin to decapitated stems could inhibit bud growth. We repeated this experiment and examined the patterns of protein expression in buds 6 and 24 hours after adding auxin. After 24 hours, the protein content was identical to that of dormant buds on intact plants, as expected. Six hours after addition of auxin, however, buds contained increased levels of growth-specific proteins and decreased levels of dormancy-specific proteins (Stafstrom and Sussex, 1988). Evidently, these buds were making progress toward the growing state before reverting to dormancy. Such cryptic growth/dormancy cycles could not be detected by measuring bud growth directly.

GENE EXPRESSION IN AXILLARY BUDS

The steady-state expression of mRNAs corresponding to several cDNA clones was studied by RNA gel blotting (Table 1). We isolated two histone genes and two ribosomal protein genes from a growing bud library and obtained additional clones from other labs. An *in situ* analysis of histone expression in tomato showed: a) that a subset of cells in growing buds contained this message; and b) that the message was absent from all cells in dormant buds (Koning et al., 1991). These results were interpreted to indicate that cells in dormant buds were arrested at some phase of the cell cycle other than S-phase and that cells of growing buds were dividing asynchronously. At what point in the cell cycle are cells in dormant pea buds arrested? Increased expression of the histone genes nine hours after decapitation (Table 1) suggests a similar delay in the onset of DNA synthesis, which means that cells would be arrested in G_1 or at the G_2/M boundary. Nagao and Rubinstein (1976) found that cells in all parts of Alaska pea buds entered mitosis 12 hours after decapitation. It is difficult to reconcile these two sets of data with cells being arrested at a single point in the cell cycle. In sucrose-starved root tips, cells are arrested in both G_1 and G_2 (Van't Hof, 1973). We recently cloned a gene that is homologous to MAP kinase. MAP kinases from *Xenopus* and mammals are activated during mitosis by phosphorylation on tyrosine (Ferrell et al., 1991; Boulton et al., 1991). We are using the expression of histones and MAP kinase to determine how the cell cycle is regulated during cycles of growth and dormancy in pea buds (Stafstrom and Anderson, in preparation; Devitt and Stafstrom, unpublished).

Stoichiometric amounts of ribosomal proteins and rRNAs are needed to build a complete ribosome. Generally, the complete set of approximately 80 ribosomal proteins accumulates coordinately, although the molecular mechanisms that account for this coordination vary among species (Gantt and Key, 1985; Mager, 1988; Amaldi et al., 1989). It was surprising, therefore, that the rate and extent of accumulation of pGB8/RPL27 were much greater than those of pGB6/RPL34. Increased expression of ribosomal components in buds would appear to be

Table 1. Relative expression of cloned genes in axillary buds, shoot apices (S), root apices (R) and fully-expanded leaves (L).

clone/identity	Lg Buds: hours after decap							Sm Buds: days after decap						S	R	L
	0	3	6	9	12	18	24	2	3	4	5	6	5+1			
pGB1/histone H2A	1	1	1	4	5	8	8	8	3	1	1	1	7	8	2	0
pGB6/rib. prot. L34	1	3	4	-	-	-	6	-	-	-	-	-	-	-	-	-
pGB8/rib. prot. L27	1	9	16	30	32	25	23	12	2	1	1	1	12	18	5	<1
pGB21/histone H4	1	1	1	5	8	8	6	6	2	1	1	1	6	8	2	0
pEA207/lectin	1	1.5	1	1.5	1	1	1	1	1.5	1	1.5	<1	1	2	0	0
pIAA4/5	1	-	2	-	-	-	23	-	-	2	-	-	-	-	-	-
pIAA6	1	-	2	-	-	-	4	-	-	1	-	-	-	-	-	-

necessary only for growth, since both dormant and growing buds synthesize proteins at similar rates (Strafstrom and Sussex, 1988).

A non-seed lectin gene corresponding to clone pEA207 is expressed at much higher levels in terminal buds than in leaves (Dobres and Thompson, 1989). Since this mRNA is present at similar levels in buds at all stages of development, its expression appears to be regulated by organ identity.

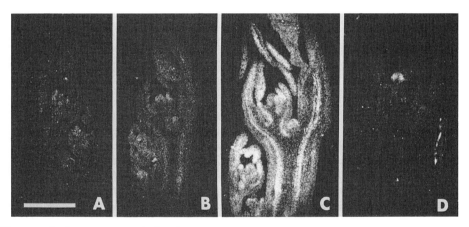

Figure 3. *In situ* expression of pGB8/RPL27 in axillary buds. Buds were from: A) intact plants; B and C) D->G buds, 1 hr or 6 hr after decapitation, respectively; and D) G->D small buds, 5 days after decapitation. Bar = 0.5 mm. For details, see Stafstrom and Sussex, 1992. Copyright American Society of Plant Physiologists, used with permission.

The rapid and considerable increase in expression of pGB8/RPL27 make it an excellent growth-specific marker. *In situ* hybridization analysis showed that RPL27 mRNA increases in all areas of the bud within one hour of decapitation (Fig. 3A and 3B). This result indicates that each bud cell is competent to respond to growth-promoting signals from the plant. The absence of a proximal-distal gradient of pGB8 expression suggests that these signals must diffuse or be transported rapidly to all cells of the bud. Six hours after decapitation, mRNA is present at maximal levels throughout the bud (Fig. 3C).

RNA gel blots indicated that RPL27 was reduced to dormant bud levels in G->D transition buds (i.e., small buds 4 to 5 days after decapitation; Table 1). However, *in situ* analysis indicated that the apical meristems of these buds contained growing bud levels of this message (Fig. 3D). Therefore, cells of the meristem respond to growth-inhibiting signals from the plant differently than neighboring cells. This cell autonomous pattern of expression could not have been predicted from RNA gel blots. These results demonstrate another way in which the G->D and the D->G transitions can be distinguished.

HORMONAL REGULATION

All five "classical" plant hormones, together with light, temperature, mineral nutrition and other factors, have been implicated in regulating bud growth (reviewed in: Phillips, 1975; Hillman, 1984; Tamas, 1988; Cline, 1991). Thimann and Skoog (1934) demonstrated that bud growth on decapitated plants could be inhibited by applying auxin in lanolin to the cut stems. Later, it was shown that that high cytokinin-to-auxin ratios could promote shoot formation *in vitro* (Skoog and Miller, 1957) and that kinetin, a cytokinin, could promote axillary bud growth despite the presence of auxin (Wickson and Thimann, 1958). Applying auxin in lanolin to decapitated plants promoted the synthesis of dormancy-specific proteins and applying kinetin directly to buds of intact plants promoted bud growth and the synthesis of growth-specific proteins (Stafstrom and Sussex, 1988). For our experiments and those of Thimann and Skoog, auxin at a concentration of 1% ($\approx 6 \times 10^{-2}$M in aqueous solution) was necessary to inhibit bud growth completely. Abnormal growth of the stem under these conditions suggested an ethylene effect (Eisinger, 1983), so it was necessary to demonstrate that bud inhibition was in fact due to auxin. We found that bud growth on cultured stem segments could be inhibited completely by $\geq 10^{-5}$M IAA and that these buds did not express pGB8/RPL27 (Stafstrom and Sussex, 1992). When stems were exposed to $\leq 10^{-6}$M IAA, the attached buds grew and expressed this message. Therefore, bud growth can be regulated by physiologically relevant concentrations of auxin.

Genetic mutants and transgenic plants with altered levels of, or sensitivity to, auxin and cytokinin corroborate the importance of these hormones in regulating bud development (Klee and Estelle, 1991). The *axr-1* mutant of *Arabidopsis thaliana* is insensitive to endogenous and exogenous auxin, and it has a bushy form indicative of weak apical dominance (Estelle and Somerville, 1987). The *trp1-1* mutant of *Arabidopsis* is tryptophan-requiring auxotroph and is therefore deficient in auxin synthesis; it too is bushy (Last and Fink, 1988). Conversely, over-expression of auxin synthesis genes derived from *Agrobacterium* in transgenic *Petunia* plants resulted in abnormally strong apical dominance (Klee et al., 1987). The *ipt* gene from *Agrobacterium* encodes a cytokinin biosynthetic enzyme. Enhanced bud development and branching occurred in transgenic tobacco plants that over-expressed this gene (Medford et al., 1989). In the studies cited above, all cells in the mutant or transgenic plants were genetically identical. Recently, transgenic tobacco plants were generated in which the *ipt* gene was interrupted by an Ac transposable element (Estruch, et al., 1991). The plants were genetic mosaics with both normal leaves and leaves containing adventious buds; the latter correlated with a reconstituted *ipt* gene, and high levels of *ipt* mRNA and cytokinins. Very abnormal, teratoma-like buds contained *ipt* mRNA whereas normal-looking buds on the same leaves did not. The absence of a cell autonomous effect indicates that cytokinins could be transported locally.

A direct role for auxin in apical dominance, as suggested by Thimann and Skoog (1934), would require that auxin content be higher in dormant buds than in growing buds. This model is difficult to reconcile with the fact that terminal buds synthesize high levels of auxin and yet grow at the same time. Based on the known polarity of auxin transport, Snow (1937) challenged the "direct theory" in an elegant set of experiments. He generated "W" shaped *Vicia faba* plants by bisecting the primary root, cotyledons and epicotyl up to the first node. When these plants were decapitated, two equal shoots developed from the cotyledonary buds. If one shoot was decapitated, buds on that shoot would not grow unless the other shoot was decapitated as well. Therefore, an inhibitor needed to travel down one shoot, up the epicotyl, down the epicotyl and then up the second shoot. It was known even then that auxin transport in the shoot is almost exclusively in the basipetal direction. More recently, Morris (1977) showed that the addition of radiolabeled auxin to one decapitated shoot on a "W" plant could inhibit bud growth on the other decapitated shoot, but no radioactive compounds reached the inhibited shoot or bud.

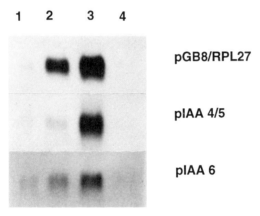

Figure 4. RNA gel blot analysis of gene expression in axillary buds. Polyadenylated RNA (0.5 µg) was isolated from: 1) D buds; 2) D->G buds, 6 hr after decapitation; 3) G buds; and 4) G->D small buds, 5 days after decapitation. The RNA was separated by formaldehyde gel electrophoresis, transferred to Nytran and probed for expression of the indicated clones.

One prediction of the indirect theory of auxin inhibition is that growing axillary buds, like growing terminal buds, should contain high levels of auxin. Recently, GC-MS analysis was used to show that auxin content in *Phaseolus* buds increases about five-fold four hours after decapitation and remains at that level through 24 hours (Gocal et al., 1991). However, a second lab using the same material and similar techniques observed no change in auxin content (LaMotte et al., 1991). We addressed the question of auxin content in pea buds by examining the expression of pIAA4/5 and pIAA6, two auxin-responsive genes (Theologis et al., 1985). Message corresponding to both clones increased in growing buds (Figure 4 and Table 1). It might be argued that reduced expression in dormant buds was due to the absence of essential transcription factors rather than reduced levels of auxin. Nonetheless, auxin must be present in growing buds and must be synthesized within the bud, since other sources of auxin (the terminal bud) have been removed. The delay in auxin-responsive message accumulation relative to pGB8/RPL27 could imply that auxin accumulation also is delayed. We hypothesize that initiation of bud growth is independent of auxin content but that continued growth requires that buds be autonomous for auxin synthesis.

If auxin does not inhibit bud growth directly, what is its role? One possibility is that auxin in the stem promotes the synthesis of a secondary inhibitor or inhibits the synthesis (or transport) of a growth promoter. Gibberellic acids cannot initiate bud growth but, when applied together with kinetin, can enhance growth (Sachs and Thimann, 1967). Ethylene has been suggested as secondary inhibitor (Blake et al., 1983). A transgenic tomato plant that over-expressed an ethylene degrading enzyme did not branch excessively, so it would appear that ethylene does not inhibit bud development (Klee, these proceedings). Physiological responses to abscisic acid (Knox and Wareing, 1984; Tamas, 1988) and the fact that this hormone declines in growing buds (Gocal et al., 1991) make it a better candidate for the secondary inhibitor. Given the discovery of additional hormone-like molecules such as systemin, salicylic acid and oligosaccharins, it would be short-sighted to consider only the five best known growth regulators.

BUD DEVELOPMENT AND PLANT ARCHITECTURE

The architectural form of a plant is one of its most characteristic features (Tomlinson, 1983). Plant architecture is influenced both by genetic or deterministic factors and by opportunistic or plastic factors. Should a branch of a plant be lost due to herbivory or other environmental assault, "reserve meristems" can develop to replace it. The pattern of new development generally mirrors or reiterates the normal developmental pattern of the plant. The need to capture photosynthetic light is perhaps the strongest driving force in the evolution of a branched canopy (Niklas, 1986). Ecologists recognize that a plant's form is a record of its history of "foraging" for light (Hutchings, 1988). Canopy shape also can affect agricultural yield (Gifford et al., 1984).

Under controlled conditions, it is possible to identify aspects of plant form that are genetically determined. We are studing several lines of pea with altered branching patterns (Figure 5). Buds on Alaska normally do not develop into branches but this capacity is retained throughout the life of the plant. Alaska is shorter than the mutant lines because it flowers at approximately node 10 rather than node 20. Parvus forms branches at four or five nodes beneath the flowering node, a condition that we call subterminal branching. L.5237 and L.5951 were derived from Parvus. These lines branch at all nodes above node 8 or 9 (aerial branching). L.5951 also forms one or two basal branches that are nearly equal in size to the main shoot. Secondary branching does not occur in L.5237 or L.5951. Putative double mutants between these lines contain numerous secondary and tertiary branches and are consequenly extremely bushy. The subterminal branching trait of Parvus behaves as a simple recessive when crossed to Alaska and it appears to be epistatic to the aerial and basal branching traits of L.5237 and L.5951. We are mapping these traits using morphological, isozyme, RFLP and RAPD genetic markers and continuing to define interactions between these loci (Polans and Stafstrom, unpublished).

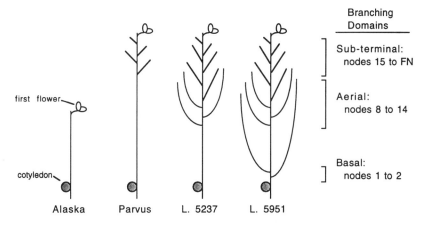

Figure 5. Schematic representation of branching patterns in genetic lines of pea. The approximate nodal positions included in each branching domain are indicated.

Can the branching phenotypes described above be accounted for by known properties of auxins and cytokinins? Auxins are synthesized predominantly in the shoot apex and transported basipetally in the shoot (acropetally in the root) and cytokinins are synthesised predominantly in the root apex and transported basipetally in the root (acropetally in the shoot). Thus, at any point along the plant axis there exists a relative and absolute level of each hormone (Figure 6). For the sake of this discussion, we will assume that buds will develop into shoots whenever the concentration of cytokinin is greater than that of auxin (bud positions below the thick horizontal line). The situation shown in Figure 6A and 6B might represent Alaska, where there is no bud development. Figure 6C shows a plant in which two basal buds have developed. This growth could be the result of decreased synthesis of auxin (6D), a decreased rate of transport or an increased rate of degradation of auxin (6E), or an increased rate of transport or a decreased rate of degradation of cytokinin (6F). Figure 6G depicts a situation in which basal buds are less sensitive to auxin or more sensitive to cytokinins. These types of models could account for basal branching in L.5951, but they cannot account for aerial and subterminal branching in any of the mutants.

The simple hormone gradient model described above could be improved greatly by considering two additional factors. First, do hormone gradients change during the course of development? And second, does the sensitivity of buds to these hormones vary as a function their ontogeny or their position on the plant axis? Subterminal branching probably results from a combination of such developmental factors. As plants flower, their terminal meristems begin to senesce and cease producing auxin. The buds nearest to the terminal buds would experience reduced auxin levels first and then begin to grow. Auxin produced by these growing buds and by developing fruits would inhibit bud growth at lower nodes. If Alaska pea plants are decapitated above node 4, buds at node 2 invariably outgrow all others (Husain and Linck, 1966). This is further evidence for an inhibitory factor moving acropetally in the shoot. Node 2 buds are larger and contain more leaf primordia than those at other nodes (Gould et al., 1987), which would allow them to grow and synthesize auxin more quickly than other buds. The same reasoning could apply to L.5951, where the growth of basal branches would inhibit branch formation at several higher nodes. These and other patterns of branch development probably result from differential sensitivity to hormones, an area of research that has received renewed interest (Palme et al., 1991; Trewavas and Gilroy, 1991).

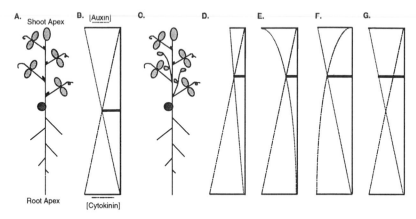

Figure 6. Model for the establishment of hormone gradients along the plant axis and their effects on branch development. See text for details.

SUMMARY AND PROSPECTS

Growth is fundamental to plant survival, but excessive growth could waste resources and be detrimental to continued survival. Apical dominance is a physiological mechanism that allows "reserve meristems" to develop only when they are needed. Under appropriate conditions, axillary bud meristems can be stimulated not only to grow, but to cease growing and, subsequently, to grow again. The ability of buds to grow so rapidly following decapitation may be due in part to their high level of metabolic activity. By studying the expression of molecular markers, we were able to detect "cryptic" cycles of growth and dormancy. It is likely that buds are constantly apprised of the health and vigor of the plant by monitoring hormone and nutrient levels, and respond accordingly.

Each stable and transition state of bud development is marked by a distinct pattern of gene expression. Understanding these states is relevant to other physiological processes. For example, winter buds of woody perennials undergo seasonal cycles of growth and dormancy. The environmental factors that regulate entry into and exit from winter dormancy are distinct from those that regulate bud development on herbaceous plants (Nooden and Weber, 1978; Powell, 1988). Nonetheless, it is likely that aspects of bud dormancy in these systems will be shared. Senescence is an irreversible developmental state that accompanies terminal differentiation (Thimann, 1987). It seems reasonable that bud dormancy, defined in molecular terms, would be an intermediate step toward bud senescence.

The growth-specific genes that we have studied have been shown to be associated with growth in other systems. The nature of dormancy-specific gene expression is more difficult to predict. Regardless, two important insights can guide us: 1) dormant buds are metabolically active, so dormancy is likely to result from active expression of dormancy-promoting genes; and 2) the cell cycle is arrested in dormant buds, so regulation of dormancy might be intimately related to regulation of the cell cycle.

Acknowledgements: I thank Ian Sussex for helping to nurture this research and Michelle Devitt, David Anderson and Sarah Murfey for helping to continue it. Sakis Theologis provided the pIAA clones and Michael Dobres provided pEA207. The branching lines described in this report were obtained from the Nordic Gene Bank, Lund, Sweden.

REFERENCES

Amaldi, F., Bozzoni, I., Beccari, E., and Pierandrei-Amaldi, P. (1989). Expression of ribosomal protein genes and regulation of ribosome biosynthesis in *Xenopus* development. Trends Biochem. Sci. **14**:175-178.

Bernier, G. (1988). The control of floral evocation and morphogenesis. Annu. Rev. Plant Physiol. Plant. Mol. Biol. **39**:175-219.

Blake, T.J., Reid, D.M., and Rood, S.B. (1983). Ethylene, indoleacetic acid and apical dominance in peas: a reappraisal. Physiol. Plant. **59**:481-487.

Boulton, T.G., Nye, S.H., Robbins, D.J., Ip, N.Y., Radziejewska, E., Morgenbesser, S.D., DePinho, R.A., Panayotatos, G.D., Cobb, M.H., and Yancopoulos, G.D. (1991). ERKs: A family of protein-serine/threonine kinases that are activated and tyrosine phosphorylated in response to insulin and NGF. Cell **65**:663-675.

Cline, M.G. (1991). Apical dominance. Bot. Rev. **57**:318-358.

Cutter, E.G. (1968). Morphogenesis and developmental potentialities of unequal buds. Phytomorphol. **17**:437-445.

Dobres, M.S., and Thompson, W.F. (1989). A developmentally regulated bud specific transcript is pea has sequence similarity to seed lectins. Plant Physiol. **89**:833-838.

Eisinger, W (1983). Regulation of pea internode expansion by ethylene. Annu. Rev. Plant Physiol. **34**:225-240.

Estelle, M.A., and Somerville, C.R. (1987). Auxin resistant mutants of *Arabidopsis* with altered morphology. Mol. Gen. Genet. **206**:200-206.

Estruch, J.J., Prinsen, E., Van Onckelen, H., Schell, J., and Spena, A. (1991). Viviparous leaves produced by somatic activation of an inactive cytokinin-synthesizing gene. Science **254**:1364-1367.

Ferrell, J.E., Wu, M., Gerhart, J.C., and Martin, S.G. (1991). Cell cycle tyrosine phosphorylation of $p34^{cdc2}$ and a microtubule-associated protein kinase homolog in Xenopus oocytes and eggs. Mol. Cell. Biol. **11**:1965-1971.

Gantt, S.J., and Key, J.L. (1985). Coordinate expression of ribosomal protein mRNAs following auxin treatment of soybean hypocotyls. J. Biol. Chem. **260**:61756181.

Gifford, R.M., Thorne, J.H., Hitz, W.D., and Giaquinta, R.T. (1984). Crop productivity and photoassimilate partitioning. Science **225**:801-808.

Gocal, G.F.W., Pharis, R.P., Yeung, E.C., and Pearce, D. (1991). Changes after decapitation of indole-3-acetic acid and abscisic acid in the larger axillary bud of *Phaseolus vulgaris* L. cv Tender Green. Plant Physiol. **95**:344-350.

Gould, K.S., Cutter, E.G., Young, J.P.W., and Charlton, W.A. (1987). Position diferences in size, morphology, and in vitro performance of pea axillary buds. Can. J. Bot. **65**:406-411.

Hillman, J.R. (1984). Apical dominance. *In* Advanced Plant Physiology, MB Wilkins, ed.; Pitman, London; pp. 127-148.

Husain, S.M., and Linck, A.J. (1966). Relationship of apical dominance to the nutrient accumulation in *Pisum sativum* var. Alaska. Physiol. Plant. **19**:992-1010.

Hutchings, M.J. (1988). Differential foraging for resources and structural plasticity in plants. Trends Ecol. Evol. 3:200-204.

Klee, H.J., Horsch, R.B., Hinchee, M.A., Hein, M.B., and Hoffman, N.L. (1987). The effects of overproduction of two *Agrobacterium tumefaciens* T-DNA auxin biosynthetic gene products in transgenic petunia plants. Genes Dev. 1:86-96.

Klee, H.J., and Estelle, M.A. (1991). Molecular genetic approaches to plant hormone biology. Annu. Rev. Plant Physiol. Plant Mol. Biol. 42:529-551.

Koning A.J., Tanimoto, E.Y., Kiehne, K., Rost, T., and Comai, L. (1991). Cell specific expression of plant histone H2A genes. Plant Cell 3:657-665.

Knox, P., and Wareing, P.F. (1984). Apical dominance in *Phaseolus vulgaris* L.: The possible roles of abscisic acid and indole-3-acetic acid. J. Exp. Bot. 35:239-244.

LaMotte, C.E., Li, X., and Cloud, N. (1991). The results of a 1977 paper showing an increased IAA level in bean axillary buds, as measured by GCMS 24 hr after decapitation, cannot be repeated. Plant Phys. 96:76s.

Lang, G.A. (1987). Dormancy: A new universal terminology. Hort. Sci. 22:817-820.

Last, R.L., and Fink, G.R. (1988). Tryptophan-requiring mutants of the plant *Arabidopsis thaliana*. Science 240:305-310.

Mager W.H. (1988). Control of ribosomal protein gene expression. Biochim Biophys Acta 949:1-15.

Medford, J.I.,Horgan, R., El-Sawi, Z., and Klee, H.J. (1989). Alterations of endogenous cytokinins in transgenic plants using a chimeric isopentenyl transferase gene. Plant Cell 1:403-413.

Morris D.A. (1977). Transport of exogenous auxin in two-branched pea seedlings (*Pisum sativum* L.). Planta 136:91-96.

Nagao, M.A., and Rubinstein, B. (1976). Early events associated with lateral bud growth of *Pisum sativum* L. Bot. Gaz. 137:39-44.

Niklas, K.J. (1986). Computer-simulated plant evolution. Sci. Am. (March) 78-86.

Nooden, L.D., and Weber, J.A. (1978). Environmental and hormonal control of dormancy in terminal buds of plants. *In* Dormancy and Developmental Arrest, M.E. Clutter, ed.; Academic Press, New York; pp221-268.

Palme, K., Hesse, T., Moore, I., Campos, N., Feldwisch, J., Garbers, C., Hesse, F., and Schell, J. (1991). Hormonal modulation of plant growth: the role of auxin perception. Mech. Dev. 33:97-106.

Phillips, I.D.J. (1975). Apical dominance. Annu. Rev. Plant Physiol. 26:341-367.

Powell, L.E. (1988). The hormonal control of bud and seed dormancy in woody plants. *In* Plant Hormones and Their Role in Plant Growth and Development, P.J. Davies, ed.; Kluwer Academic, Dordrecht, The Netherlands; pp 539-552.

Sachs, T., and Thimann, K.V. (1967). The role of auxins and cytokinins in the release of buds from apical dominance. Am. J. Bot. 54:136-144.

Skoog, F., and Miller, C.O. (1957). Chemical regulation of growth and organ formation in plant tissues cultured *in vitro*. Symp. Soc. Exp. Biol. 11:118-131.

Snow, R. (1937). On the nature of correlative inhibition. New Phytol. 36:283-300.

Stafstrom, J.P., and Sussex, I.M. (1988). Patterns of protein synthesis in dormant and growing vegetative buds of pea. Planta 176:497-505.

Stafstrom, J.P., and Sussex, I.M. (1992). Expression of a ribosomal protein gene in axillary buds of pea. Plant Physiol. In Press.

Sussex, I.M. (1989). Developmental programming of the shoot meristem. Cell 56:225-229.

Tamas, I.A. (1988). Hormonal regulation of apical dominance. *In* PJ Davies, ed., Plant Hormones and Their Role in Plant Growth and Development; Kluwer, Dordrecht, The Netherlands; pp393-410.

Theologis, A., Huynh, T.V., and Davis, R.W. (1985). Rapid induction of specific mRNAs by auxin in pea epicotyl tissue. J. Mol. Biol. 183:53-68.

Thimann, K.V. (1987). Plant senescence: A proposed integration of the constituent processes. *In* Plant Senescence: Its Biochemistry and Physiology, ASPP; pp 1-19.

Thimann, K.V., and Skoog, F. (1934). On the inhibition of bud development and other functions of growth substance in Vicia faba. Proc. R. Soc. London Ser. B 114:317-339.

Tomlinson, P.B. (1983). Tree architecture. Amer. Sci. 71:141-149.

Trewavas, A., and Gilroy, S. (1991). Signal transduction in plant cells. Trends Genet. 7:356-361.

Van't Hof, J. (1973). The regulation of cell division in higher plants. *In* Basic Mechanisms in Plant Morphogenesis, Brookhaven Symp. 25:152-165.

Wickson, M., and Thimann, K.V. (1958). The antagonism of auxin and kinetin in apical dominance. Physiol. Plant. 11:118-131.

ELICITATION OF ORGANIZED PIGMENTATION PATTERNS BY A CHALCONE SYNTHASE TRANSGENE

Richard A. Jorgensen

Department of Environmental Horticulture and
Department of Vegetable Crops
University of California-Davis
Davis, California 95616-8587

ANTHOCYANINS AS MARKERS FOR DEVELOPMENTAL PATTERNS

The problems of determination and elaboration of developmental patterns are more accessible in the case of visible patterns such as anthocyanin pigmentation patterns. Anthocyanin pigments offer the added advantage that they are completely dispensible in plant growth and development, which means that manipulation of anthocyanin pigmentation patterns generally has no deleterious or pleiotropic effects on the plant. Furthermore, many anthocyanin genes are known to act cell autonomously, so patterns of expression can visualized on a cell by cell basis. Coen et al. (1988) have discussed the basic principles behind flower color patterns with emphasis on *Antirrhinum majus*. Here I will focus on flower color patterns in *Petunia hybrida* and explain how the use of transgenes to manipulate patterns provides a tool with the potential to improve our understanding of how patterns are determined and elaborated.

MODES AND LEVELS OF PIGMENTATION PATTERN DETERMINATION

Some anthocyanin pigmentation patterns are determined by pattern-specific expression of regulatory genes which encode *trans*-acting factors for structural gene transcription. In this type of pattern determination, a prepattern of regulatory gene expression determines the expression of most or all of the structural anthocyanin genes. Examples of such regulatory genes in petunia are the *An1* and *An2* genes which regulate all genes downstream of the dihydroflavonol reductase step (Gerats *et al.*, 1983). This type of mechanism is common in the determination of patterns which follow tissue-specific boundaries. Thus, the pattern results simply from control of regulatory gene expression by factors which are expressed according to tissue type determination. Patterns within a tissue type may also be determined in this fashion if the presence or abundance of a regulatory factor varies spatially within the tissue. How the prepattern of regulatory gene expression is determined is unknown, but centrally important to understanding determination.

Pigmentation patterns may also be determined by the control of a single structural gene in a pathway such that, although all genes in the pathway have been turned on according to tissue-specific signals, one gene is subject to an additional level of pattern-specific control. This use of multiple levels of pattern determination would seem to provide for greater potinetial diversity in pattern form. Examples of a second level of pattern determination in petunia are the familiar "star" and "picotee" patterns of garden petunias. These patterns are due to control of chalcone synthase expression in flower petals, resulting in white sectors in which chalcone synthase is not expressed and the remainder of the pathway is expressed (Mol et al., 1983).

Chalcone synthase (CHS) acts cell autonomously, and the white sectors of patterns are non-clonal, so the star and picotee patterns must be determined by intercellular communication. Importantly, sectors are coincident in the upper and lower epidermal layers of the corolla, indicating that signals move perpendicularly to the surface of the petals. The mechanisms behind chalcone synthase regulation and intercellular signalling are unknown; however, nuclear run on transcription experiments on isolated nuclei suggest that regulation is at the post-transcriptional level (van der Meer, 1991). Genetic analysis of star patterns has indicated that the pattern phenotype is controlled by multiple genes (T. Holtrop, A. Gerats, personal communications).

PATTERN ELICITATION BY A CHALCONE SYNTHASE TRANSGENE

Fortuitously, a system has been discovered in which patterns similar to the star and picotee petunias can be elicited (Napoli et al., 1990; van der Krol et al., 1990a,b). This system is based on the introduction of a chalcone synthase transgene into petunia inbred lines which are otherwise uniformly and stably pigmented. It was found that the transgene elicited a variety of non-clonal patterns which are due to the co-suppression of both the endogenous chalcone synthase gene and the transgene in the white regions of the flower.

PATTERN PERTURBATION THROUGH TRANSGENE ALLELISM

Both interestingly and usefully, the CHS transgene occasionally undergoes spontaneous epimutation to produce distinct new patterns. These new epiallelic forms of the transgene locus are reversible. They are also epimutable to produce new alleles eliciting other patterns. Progeny testing of branches in which somatic epimutations have occurred reveals that many of these events are germinally heritable. The molecular mechanism of transgene epimutability is unknown, but may involve DNA methylation, chromatin organization, and/or reversible DNA rearrangements.

This heritability was unexpected in view of the fact that the pattern phenotype is elaborated by control of chalcone synthase expression in the L1-derived epidermis, while gamete-producing cells are L2-derived, i.e., from a clonally distinct cell lineage. This suggests that pattern elaboration in the epidermis is controlled by signals from the mesophyll layer, consistent with the observation that patterns are coincident in the upper and lower epidermal layers. Another possible interpretation is that transgene epimutation occurs simultaneously in the two cell layers. This possibility is perhaps not as unlikely as it might seem since some epimutations do appear to occur non-randomly in non-clonal clusters, especially in aging plants. Such behavior of epimutations in maize was reported by McClintock (1967).

INTERCELLULAR COMMUNICATION

Which cells are responsible for pattern determination? Inspection of the full range of patterns elicited by various transgene loci and alleles of one locus suggests that patterns are determined by cells at the edges of petals and/or cells in or near the vasculature of the petal. In addition, subsets of cells within these two types may be responsible for pattern determination. For instance, petal edge cells which comprise the fused junction between petals may act to express pattern while other edge cells not involved in petal fusion do not, resulting in one type of "star" pattern (examples shown in Figure 1). The reciprocal pattern is also possible, yielding the "picotee" pattern (i.e., giving the appearance of a white ring surrounding a pigmented center) produced by breeders. Also, cells at the midrib may determine another "star" pattern, or many veins may determine a "watershed" pattern. Sometimes, the effect is most pronounced at the base of the flower, other times at its outer end, indicating an influence of the corolla base or tip. Also, the effect is typically more pronounced in the lower two petals than in the upper three, indicating an influence related to the zygomorphy of the petunia flower. It is intriguing that different epialleles of a single transgene locus can produce all these pattern types, except the picotee type, which nevertheless has been observed to be elicited by other transgene loci.

The flowers shown in Figure 1 all derive from the same plant, illustrating the variability of patterns within an individual. The star-like pattern of these flowers appears to be the result of cells at the junction between adjacent petals. The fusion of adjacent petals in solanaceous flowers is known to be the result of redetermination of L1-derived epidermal cells into mesophyll cells at the time of fusion. Typically, epidermal cells from both adjacent petal initials contribute to these new mesophyll cells. The patterns in Figure 1 appear to be determined by either these fusion zone mesophyll cells or the fusion zone epidermal cells or both. It is interesting to note that size (length and width) of a white stripe is always the same in both petals, indicating that the pattern is determined cooperatively between the two adjacent petals, perhaps by signals originating in the fusion zone cells.

Figure 1. Flowers from a single plant carrying a CHS transgene.

Useful information on pattern elaboration may be obtained from observation of patterns in lines which are somatically unstable, producing epigenetic variants, even within flowers. Epigenetic variants can have two origins, clonal or inductive (cooperative). An example of inductive variants was illustrated above in the flowers of Figure 1. An example of a putative clonal sector is illustrated in the three upper panels of Figure 2. Most large clonal sectors recognizable in petunia flowers have their boundaries at the junction between petals and to a lesser extent at the midrib. This is because in the developing floral meristem petal initiation involves a small proportion of cells in the petal whorl and most of these give rise to the midrib. Thus, large clonal sectors are likely have their boundaries in regions of the meristem which do not give rise to cells of the petal between its midrib and its edge. Flowers with apparent clonal sectors have sectors that run down the tube of the flower all the way to its base. An example of an apparent clonal sector is shown in Figure 2, in which the upper left and upper center panels show the first and second flowers produced on a plant and the upper right panel shows a side view of the first flower demonstrating that the sector extends to the corolla base. It is not known yet whether such sectors are due to clonal inheritance; however, they do provide an illustratration of what a clonal sector within L2 cells would look like in petunia flowers. Of course, the other possible origin of such sectors is an inductive "reprogramming" of a meristem or an organ initial (or multiple adjacent initials) through intercellular commnunication. Such a process would be equivalent to the reprogramming of a meristem and its constituent cells from a vegetative to a floral state.

The lower three flowers in Figure 2 illustrate the expected appearance of clonal sectors in a pattern type in which the cells of the petal fusion zone determine pattern. In Figure 1 the same pattern type was analyzed, but no apparently clonal events were observed. The line giving rise to the flowers in the lower row of Figure 2 produces the pattern of the lower left flower in most plants. This pattern is a star-like pattern, but only at the lower three petal fusion zones, unlike the pattern of Figure 1. At high frequency flowers are observed like those in the lower center and lower right panels of Figure 2.

In the flower in the lower right panel, the junction between the lower left and

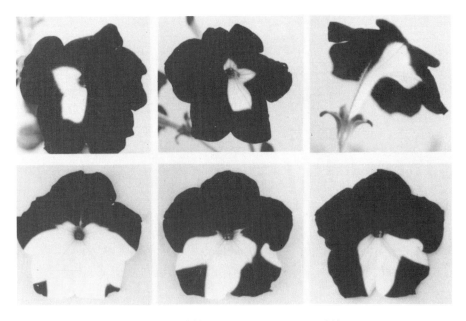

Figure 2. Flowers exhibiting putative clonal sectors within patterns.

upper left petals appears not to determine a white sector. This flower might be interpreted as being comprised of two clonal lineages: one comprised of the upper three petals and the other comprised of the lower two petals. The wide white stripe centered on the junction between the two lower petals extends to junction between to two leftmost petals, but does not cross this zone, as if the cells in the upper left petal are not responsive to the pattern determining signal from the junction between the lower petals. On the right side of the flower, the junction between the lower and upper petals appears to be the origin of a white stripe, but only the fusion zone cells near the edge of the corolla are so competent. Note also that the white zone extends only a short distance into the upper right petal at its edge, as if the upper petal is not responsive to the signal generated in the fusion zone. One interpretation is that the fusion zone near the petal edge is comprised of more cells from the lower right petal than the upper right petal and are able to detemine a white stripe in that region.

Similarly for the flower in the lower center panel of Figure 2, one could interpret it as being comprised of two clonal lineages, a non-patterned lineage in the upper three flowers and a pattern eliciting lineage in the lower two petals. The lower left/upper left petal fusion zone appears similar to the lower right/upper right fusion zone of the previously described flower. The lower right/upper right fusion zone can also be interpreted in the same manner as these, except that it appears that a smaller region of cells at the corolla edge is involved in determining the sector in that position.

An alternative interpretation of these patterns would be that each petal initial must be determined for pattern elaboration in an inductive manner. Fusion of differently determined cells from adjacent petals might result in the same interactions as for clonal lineages. It is the coincidence of most clonal lineage boundaries with petal fusion zones, as well as the diffuseness of sector boundaries, that makes it difficult to distinguish between these two possibilities.

A MODEL SYSTEM FOR PATTERN DETERMINATION?

The transgenic pattern elicitation system described here may be useful for investigating the basis of pattern determination and/or elaboration. The challenge in realizing this potential will be to understand the mechanism by which the transgene is responsive to developmental signals. If this responsiveness can be understood, then it may be possible to investigate the signal(s) to which the transgene is sensitive.

ACKNOWLEDGEMENTS

The plants discussed here were derived from lines provided by DNA Plant Technology Corporation and Florigene BV. I would like to thank John Bedbrook for making it possible to obtain these lines, Nancy Kerk, Ian Sussex, and Dick Flavell for stimulating discussions on pattern elaboration, Xiao-ming Sha and Anne Marie Napoli for assistance with plant tissue and greenhouse culture, John Yoder, Jim Harding, and Carolyn Napoli for providing research facilities, and Carolyn Napoli for support and encouragement throughout the project. This research was funded in part by a small grant from the USDA NRICRG program, as well as through the author's own personal resources.

REFERENCES

Coen, E.S., Almeida, J., Robbins, T.R., Hudson, A., and Carpenter, R., 1988, Molecular analysis of genes determining spatial patterns in *Antirrhinum majus, in*: "Temporal and Spatial Regulation of Plant Genes," D.P.S. Verma and R.B. Goldberg, eds., Springer-Verlag, Wien.

Gerats, A.G.M., Wallroth, M., Donker-Koopman, W., Groot, S.P.C., and Schram, A.W., 1983, The genetic control of the enzyme UDP-Glucose:3-O-flavonoid-glucosyltransferase in flowers of Petunia hybrida, *Theor. Appl. Genet.* 65:349.

McClintock, B., 1967, Genetic systems regulating gene expression during development, *Devel. Biol. Suppl.* 1:84.

Mol, J.N., Schram, A.W., de Vlaming, P., Gerats, A.G., Kreuzaler, F., Hahlbrock, K., Reif, H.J. Reif, and Veltkamp, E., 1983, Regulation of flavonoid gene expression in *Petunia hybrida*: description and partial characterization of a conditional mutant in chalcone synthase gene expression, *Mol. Gen. Genet.* 192:424.

Napoli, C., Lemieux, C., and Jorgensen, R., 1990, Introduction of a chimeric chalcone synthase gene into petunia results in reversible co-suppression of homologous genes in trans, *Plant Cell* 2:279.

van der Krol, A.R., Mur, L.A., Beld, M., Mol, J.N.M., and Stuitje, A.R., 1990a, Flavonoid genes in petunia: addition of a limited number of gene copies may lead to a suppression of gene expression, *Plant Cell* 2:291.

van der Krol, A.R., Mur, L.A., de Lange, P., Gerats, A.G., Mol, J.N., and Stuitje, A.R., 1990b, Antisense chalcone synthase genes in petunia: visualization of variable transgene expression, *Mol. Gen. Genet.* 220:204.

van der Meer, I.M., 1991, "Regulation of flavonoid gene expression in Petunia hybrida: cis-acting elements and trans-acting factors," Ph.D. Dissertation. Vrije Universiteit te Amsterdam. Amsterdam.

PATTERN FORMATION DURING *ARABIDOPSIS* EMBRYO DEVELOPMENT

Ulrike Mayer, Thomas Berleth, Ramon A. Torres Ruiz, Simon Miséra, and Gerd Jürgens

Institut für Genetik und Mikrobiologie
Universität München
8000 Munich 19, FRG

INTRODUCTION

A developing organism is a growing population of cells that, at one stage or another, exchange information about their relative positions. Although cell communication appears to play an important role in generating the body organization, it is hard to define the components involved as well as their interactions. This is especially true of pattern formation in the developing plant embryo which produces the primary body organization as represented by the seedling.

During plant embryogenesis, cell number increases several thousand-fold, and the growing embryo undergoes a series of distinct shape changes which result from position-dependent cell activities. As the cells are immobilized by their walls, they respond to positional cues by changing mitotic rate, by orienting their planes of division or by directional expansion associated with unequal division (Lyndon, 1990). While the mechanics of cell division and expansion have been studied for many years (Lloyd, 1991), comparatively little is known about the positional regulation of these processes in the developing embryo. What are the signals, how are they generated and distributed, how are they perceived and transduced, and how do the cells respond to them? One way of addressing these questions is the genetic approach we have taken to analyze pattern formation in the *Arabidopsis* embryo. In short, relevant genes are identified on the basis of their mutant phenotypes and studied at the molecular level in order to characterize their modes of action. Here, we discuss some aspects of pattern formation as revealed by mutant phenotypes. To provide a starting point, we briefly describe the body organization of the seedling and how it develops in the embryo.

DEVELOPMENT OF THE PRIMARY BODY ORGANIZATION

The seedling represents the primary body organization as established in the embryo before the shoot and root meristems take over to produce the adult plant body using the pre-existing organization as reference. Two patterns are superimposed in the seedling (Fig. 1). An apical-basal pattern is arranged along the single axis of polarity which is to become the main axis of the plant. From top to bottom, the following elements can be distinguished: epicotyl including the shoot meristem, cotyledons, hypocotyl, and root including the root meristem and root cap. The second pattern runs perpendicular to the axis of polarity, and this radial pattern comprises the main tissues, such as epidermis, ground tissue (cortex) and vascular tissue, as seen in the hypocotyl (Fig. 1). The radial pattern of the root includes additional elements, e.g endodermis and pericycle (Schiefelbein and Benfey, 1991). Finally, the spatial arrangement of the cotyledons which marks a bilaterally symmetric organization may be viewed as a modification of the radial pattern.

Figure 1. Schematic presentation of the seedling organization. The apical-basal pattern is shown on the left, the radial pattern on the right. C, cotyledons; E, epidermis; G, ground tissue; H, hypocotyl; R, root and root meristem; S, epicotyl and shoot meristem; V, vascular tissue.

The primary body organization originates during embryogenesis. Due to the relatively regular patterns of cell division in many plant species, it has been possible to trace, by histological means, the "lineages" of seedling pattern elements back to successively earlier stages of embryogenesis. This has been done in great detail for the crucifer *Brassica napus* (Tykarska, 1976, 1979), and the patterns of cell division described for the early embryo appear to be identical in other crucifers such as *Arabidopsis thaliana* (Mansfield and Briarty, 1991; Jürgens and Mayer, 1992) and *Capsella bursa-pastoris* (Schulz and Jensen, 1968). Such cell "lineages" do not imply commitment of cells, or groups of cells, to specific fates but could reflect position-dependent cell activities. Nonetheless, knowing how the pattern normally develops is very helpful in analyzing mutant phenotypes.

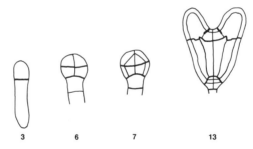

Figure 2. Development of the apical-basal pattern in the embryo. Bold lines mark subdivisions of the axis. Numbers indicate stages of embryogenesis according to Jürgens and Mayer (1992).

By morphological criteria, the apical-basal pattern develops in steps (Fig. 2). Following fertilization, the polarized egg cell expands in the apical-basal axis before the first cell division takes place (see also Webb and Gunning, 1991). This asymmetric division divides the zygote into a small apical cell and a large basal cell. The apical cell will give rise to the embryo, except that part of the root meristem, together with the extra-embryonic suspensor, will be produced by the basal cell. By the octant stage (stage 6 of Jürgens and Mayer, 1992), the apical cell has been partitioned into two nearly equal-sized regions. The upper region will later be subdivided into the epicotyl, including the shoot meristem, and the cotyledons while the lower region will give rise to the hypocotyl, the embryonic root and initials of the root meristem. These later subdivisions become apparent in the heart-shaped embryo. The other daughter cell of the zygote, the basal cell, divides transversely and, by the dermatogen stage (Mansfield and Briarty, 1991; stage 7), has produced a file of 7 - 9 cells.

The uppermost of these cells is the hypophysis which, by asymmetric division, generates a small lens-shaped cell abutting the rest of the embryo, and a larger cell connected with the suspensor. The lens-shaped cell will give rise to the four central cells of the root meristem, which are present from the mid-heart stage (stage 13) on and may correspond to the quiescent center. The other daughter cell will produce the central portion of the root cap, including initials and cell layers.

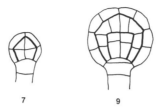

Figure 3. Development of the radial pattern in the embryo. The bold lines separate the tissue primordia. Numbers indicate stages of embryogenesis.

The basic elements of the radial pattern become apparent in a comparatively short time interval, from the dermatogen stage (stage 7) to the mid-globular stage (stage 9; Fig. 3). The first element to appear is the primordium of the epidermis which originates as the outer layer of daughter cells produced by tangential cell divisions in the octant-stage embryo. Subsequently, the integrity of the epidermal cell layer is maintained by divisions along the surface. The inner cells of the dermatogen-stage embryo divide in parallel to the apical-basal axis, producing elongate cells in the center of the embryo from which the vascular tissue of the hypocotyl originates. In the developing root, the radial pattern is elaborated from the mid-heart stage on and is essentially complete in the torpedo-shaped embryo. Above the hypocotyl, the symmetry changes from radial to bilateral during the transition from the late-globular to the early-heart stage. This change is brought about by the cotyledonary primordia which are emerging from two sites across the apical-basal axis. In summary, the body organization of the *Arabidopsis* seedling is outlined in the early embryo when less than 200 cells are present, and later refined, including the formation of the shoot and root meristems.

MUTATIONS AFFECTING THE BODY ORGANIZATION

Our search for embryonic pattern mutants was based on the assumption that mutations in genes directing the spatial organization of the developing embryo should not interfere with the completion of embryogenesis, and mutant seedlings were expected to show specific morphological abnormalities (see Jürgens et al., 1991, for discussion of the strategy used). Although about 250 putative pattern mutants were identified in this way, we have focused on the analysis of 73 mutants showing major deviations from the wild-type seedling organization. These mutants define 9 complementation groups, and each gene has its own specific mutant phenotype (Mayer et al., 1991). Considering the average allele frequency, the mutant phenotypes are likely to result from loss-of-function mutations and, therefore, indicate how the developing system responds to the inactivation of individual components.

Mutant phenotypes revealing early patterning events

Three different aspects of the body organization are affected independently of each other in different mutants: apical-basal pattern, radial pattern and shape. Mutations in four genes delete different regions of the apical-basal pattern (Fig. 4). Apical structures, including cotyledons and shoot meristem, are missing, or strongly reduced, in *gurke* mutant seedlings, *fackel* mutant seedlings seem to lack the hypocotyl, and mutations in the *monopteros* gene delete both the hypocotyl and the root. In contrast to these fairly homogeneous phenotypes, all mutant alleles of *gnom* produce two different phenotypes: a majority of mutant seedlings lack apical and basal structures while others show no axial polarity. In the latter case, however, the tissues of the radial pattern are present although the cells of the vascular tissue are not interconnected to form strands, suggesting that apical-basal and radial patterns originate separately. Developing mutant embryos were found to deviate from normal, and the relative positions of defects in the heart-shaped embryo correspond well with the positions of the deleted seedling structures (Fig. 4), in agreement with the histological "lineage" studies mentioned above. The mutant phenotypes therefore suggest that the apical-basal axis is initially partitioned into three large regions, apical, central and basal, roughly corresponding to cotyledons and shoot meristem, hypocotyl and root, respectively.

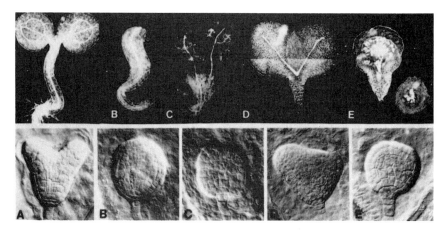

Figure 4. Phenotypes of apical-basal pattern deletion mutants at the seedling stage (upper panel) and at the early-heart stage (lower panel). A, wild-type; B, gurke; C, fackel; D, monopteros; E, gnom. For details, see text.

Mutations in two genes cause radial pattern defects. Although the mutant seedlings look grossly abnormal, presumably due to secondary effects, the early embryonic phenotypes strongly suggest that the genes are required for the formation and/or development of the epidermal primordium (Fig. 5). In *knolle* embryos, there is no outer cell layer corresponding to the epidermal primordium, and the cells are enlarged. By contrast, *keule* embryos have a distinct layer of epidermal precursor cells which, however, look abnormal from early on. We have isolated, but not yet characterized genetically, additional mutants with similar phenotypes (U. Mayer, unpublished).

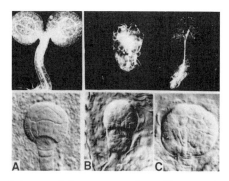

Figure 5. Phenotypes of radial pattern defect mutants at the seedling stage (upper panel) and at the early-globular stage (lower panel). A, wild-type; B, knolle; C, keule. For details, see text.

Of the three genes mutating to abnormal seedling shape phenotypes, one is particularly interesting because cell shape appears to be abnormal in developing *fass* mutant embryos from early on (not shown; see Mayer et al., 1991). Yet, the elements of both the apical-basal and the radial pattern are all present (Fig. 6). Thus, pattern formation does not require the ability of cells to undergo shape changes although characteristic shape changes are normally associated with the development of specific pattern elements such as epidermis or vascular primordium. In addition, *fass* mutant seedlings can give rise to adult plants which, however, are stout and sterile (Fig. 6; R.A. Torres Ruiz, unpubl.).

Mutant phenotypes suggesting links between embryo and meristem organization

Root and shoot meristems are established late in embryogenesis. Defects in the root meristem were comparatively easy to detect in our screen as the root meristem is active from germination on, producing a rapidly growing root, and the embryonic root (radicle) is well differentiated. Root mutants were recognized by the lack of root growth or abnormal root shape but these defects have not been analyzed in detail (for a general discussion of the genetic dissection of root development, see Schiefelbein and Benfey, 1991). By contrast, mutations affecting shoot meristem formation, are not well represented in our collection of embryonic pattern mutants, due to our screening procedure (Jürgens et al., 1991). Nonetheless, we have isolated mutants with specific apical defects, which fall into different phenotypic classes: both shoot meristem and cotyledons are altered, or one but not the other is affected.

The number of cotyledons is changed in several mutants. For example, mutations in the *häuptling* gene increase the number of cotyledons to four or more (Fig. 6). Compared to wild-type, the mutant plants later produce more leaves and these are abnormally spaced. The same correlation between number of cotyledons and spacing of leaves has been observed for other mutants, suggesting that phyllotaxis involves the cotyledons as references. We have also found a different class of mutants in which the newly-emerged mutant seedlings look like wild-type but become abnormal by the time the wild-type seedlings have started to produce leaves. For example, *reduced apical meristem* mutant seedlings with one or no leaves do not change for about two weeks until lateral meristems start to develop, eventually producing fertile flowers (Fig. 6; T. Laux and G. Jürgens, unpubl.). Thus, this class of mutations appear to affect the formation of the embryonic shoot meristem. A complementary phenotype is observed in *laterne* mutant seedlings which lack cotyledons but later produce leaves (Fig. 6; R.A. Torres Ruiz, unpubl.). In summary, the apical-defect phenotypes emphasize the significance of the seedling organization for the development of the adult plant body.

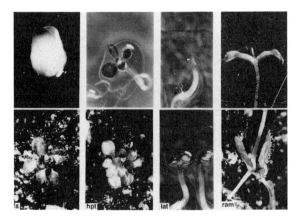

Figure 6. Phenotypes of apical-defect mutants at the seedling stage (upper panel) and during postembryonic development (lower panel). fs, fass; hpt, häuptling; ram, reduced apical meristem; lat, laterne. For details, see text.

CONCLUSIONS

The results of our search for embryonic pattern mutants demonstrate that pattern formation can be dissected genetically. The mutant phenotypes suggest that early events generate a broad outline of the body organization which includes partitioning of the apical-basal axis and formation of tissue types. Later events are confined to particular regions of the embryo, such as the apical and basal ends, and appear to bring about the organization of the meristems. It is clear from the phenotypes analyzed so far that not each and every aspect of pattern formation in the embryo has been revealed, and therefore, our continuing efforts of isolating additional pattern mutants are likely to be rewarding.

The genes that have been characterized genetically, especially those involved in the partitioning of the axis, seem to act in a position-dependent manner. What roles they play in pattern formation can be tested when molecular probes become available. Considering their mutant phenotypes, molecular analysis of these genes may eventually contribute to our understanding of cell communication in the developing plant embryo.

ACKNOWLEDGMENTS

We wish to thank Gabi Büttner for excellent technical assistance and the other members of our group for helpful comments on the manuscript.

REFERENCES

Jürgens, G., and Mayer, U., 1992, Arabidopsis, *in:* "A Colour Atlas of Developing Embryos", J. Bard, ed., Wolfe Publishing, London (in press)

Jürgens, G., Mayer, U., Torres Ruiz, R.A., Berleth, T., and Miséra, S., 1991, Genetic analysis of pattern formation in the *Arabidopsis* embryo, *Development Suppl.* 1:27

Lloyd, C.W., 1991, How does the cytoskeleton read the laws of geometry in aligning the division plane of plant cells? *Development Suppl.* 1:55

Lyndon, R.F., 1990, "Plant Development. The Cellular Basis", Unwin Hyman, London

Mansfield, S.G., and Briarty, L.G., 1991, Early embryogenesis in *Arabidopsis thaliana*. II. The developing embryo. *Can. J. Bot.* 69:461

Mayer, U., Torres Ruiz, R.A., Berleth, T., Miséra, S., and Jürgens, G., 1991, Mutations affecting body organization in the *Arabidopsis* embryo. *Nature* 353:402

Schiefelbein, J.W., and Benfey, P.N., 1991, The development of plant roots: new approaches to underground problems. *Plant Cell* 3:1147

Schulz, R., and Jensen, W.A., 1968, *Capsella* embryogenesis: The egg, zygote, and young embryo. *Am. J. Bot.* 55:807

Tykarska, T., 1976, Rape embryogenesis. I. The proembryo development. *Acta soc. bot. pol.* 45:2

Tykarska, T., 1979, Rape embryogenesis. II. Development of the embryo proper. *Acta soc. bot. pol.* 48:391

CELL COMMUNICATION AND THE COORDINATION OF DIFFERENTIATION

Judith A. Verbeke

Department of Plant Sciences
University of Arizona
Tucson, AZ 85721

INTRODUCTION

The processes of cellular and tissue differentiation are of fundamental significance to our understanding of plant growth and development, yet experimental studies in this area have been rare. During development, each cell must somehow determine its position relative to others, and must differentiate accordingly. The cells in higher organisms do not develop independently as preprogrammed units but rather are targets for various extrinsic stimuli. Groups of cells apparently communicate, and thereby set or reset specific programs of gene expression. However, almost nothing is known about the mechanisms by which cells establish their positions and subsequently give rise to appropriate cell types. Although cell and tissue differentiation must certainly be the result of selective gene expression, a critical problem has been our inability to make the connection between phenotype and genotype. At a time when our knowledge of gene structure and function seems to be increasing exponentially, we still have little knowledge about how these genes actually cause cells to differentiate and form tissues, organs, and organisms.

CHEMICAL CELL-CELL SIGNALING IN PLANTS

In higher plants especially, cell and tissue differentiation seems to depend more on the position in which particular cells find themselves than on their clonal lineage, indicating that the cell is somehow receiving signals which contain positional information. The initial event appears to be cell recognition, involving a cell-surface phenomenon thought to be dependent on the complexing of the signal molecule and a membrane-bound receptor (Knox and Clarke, 1984). Obvious recognition events occur during fertilization in lower and higher plants. They also occur between somatic cells of flowering plants during grafting and in interactions between higher plants and microorganisms. What these recognition events have in common is the fact that they are all exceptions to a basic "rule" in plant growth and development – that no new contact occurs between cells that have not arisen together as a clone after cell division because the wall usually prohibits the morphogenetic movements that are commonplace in animal development.

There are also examples of cell-cell communication in plants which do not involve new contact between cells. Utilizing fate-mapping, Sarah Hake and Mike Freeling (1986) have shown that it is the genotype of mesophyll cells and not the genotype of epidermal cells which is critical for the expression of altered leaf development in the *knotted* mutant in maize. This neomorphic mutation causes either a vast overproduction of a normal product, an altered non-competing product, or an entirely new product in the mutant leaves

(Hake et al., 1989). Whatever the case may be, their data clearly show that some signal from the mesophyll cells induces the epidermis to divide (Sinha and Hake, 1990).

Another example of chemical cell-cell signaling in plants has been studied in our laboratory. Plant epidermal cells usually display a high degree of determination, as indicated by a resistance to redifferentiate. In the absence of wounding (and usually in spite of wounding), plant epidermal cells are developmentally incompetent to redifferentiate. These cells show no redifferentiation (or grafting) response even after months of intimate contact with either wounded or non-wounded surfaces (Moore, 1984; Walker and Bruck, 1985). This stable differentiated state which is exhibited by plant epidermal cells arises early in ontogeny (Bruck and Walker, 1985). However, fusing floral organs are naturally occurring exceptions to this rule. Postgenital fusions are tissue unions which occur by adhesion of the outer epidermal cell walls of the fusing organs in the absence of protoplasmic union (Cusick, 1966). The fusion of floral organs is an important developmental event in the morphogenesis of many flowers which occurs in several families of higher plants (Boeke, 1971; Nishino, 1982; Verbeke, 1992). The course of the fusion does not vary much; the parts are forced together by growth, the epidermal cells interlock and the enclosed cuticle, if present, disappears. In the fused epidermal layers, cell divisions – which are mostly periclinal – usually occur which disturb the original cell pattern and obscure the exact place of fusion. In the region of coalescence, the identity of the originally separate epidermal layers is lost. These fusions involve a change in the developmental fate of the contacting epidermal cells.

CATHARANTHUS ROSEUS AS AN EXPERIMENTAL SYSTEM

A clear example (Boke, 1947, 1948, 1949) of postgenital fusion occurs in the developing gynoecium of *Catharanthus roseus* (Madagascar periwinkle). During normal floral ontogeny in *C. roseus*, the adaxial surfaces of the two carpels, which are originally separate, touch. Within nine hours, approximately four hundred contacting epidermal cells then convert, by redifferentiation, into parenchyma cells (Verbeke and Walker, 1985). The changes in these fusing cells are dramatic. Prefusion epidermal cells have a thin cuticle, are rectangular in section and densely cytoplasmic, and divide strictly in the anticlinal plane. Following contact, these cells still have the remnants of the cuticle, but become isodiametric and highly vacuolated and divide predominantly in the periclinal plane (Walker, 1975a, 1975b, 1975c).

We have utilized this experimental system to address basic questions about cell communication and floral development. This has proven to be a particularly useful experimental system for a variety of reasons:

• The cells which ultimately change their developmental fate are initially epidermal, which means they are external and can be subjected to manipulation with a minimum of wounding. The changes we observe, therefore, can be attributed unequivocally to the experimental conditions rather than to a wound response.

• In periwinkle, the epidermal cell redifferentiation occur within nine hours of the initial contact (Verbeke and Walker, 1985). The responses we monitor are clearly produced very soon after the initial cell contact and cell communication.

• The distinctive cytoplasm and cell shape of the epidermal cells makes it possible to identify changes as the cells redifferentiate.

• The qualitative changes which occur in the redifferentiating cells have been well described and the events are well documented (Boke, 1947, 1948, 1949; Walker, 1975a, 1975b, 1975c, 1978a, 1978b).

SUMMARY OF EXPERIMENTAL RESULTS

We have addressed the issue of how the cells might know where they are, and what to become next. That is, how do the originally epidermal cells "know" they are no longer on the outside and that they should now redifferentiate into another cell type? Some of the specific questions we have addressed include the following.

Are the Redifferentiating Cells Temporally Programmed to Redifferentiate?

Will the epidermal cells automatically change their fate when they reach a certain age/developmental state? We surgically removed one of the carpels at a prefusion height and monitored the developmental state of the epidermal cells in the remaining carpel. In the absence of the other carpel, all epidermal cells remained epidermal (Walker, 1978a); no redifferentiation occurred.

Do the Cells Perceive a Change in Pressure?

Does the mere act of one carpel surface pressing against the other cause redifferentiation? When we inserted a solid, impermeable barrier between the two carpels, all epidermal cells remained epidermal (Verbeke and Walker, 1986; Walker, 1978a); no redifferentiation occurred.

Does a Surface Recognition Event Occur or are Diffusible Factors Involved?

Do the surfaces have to physically contact one another to cause redifferentiation or are the cells somehow communicating chemically with one another? We placed barriers with pores of different sizes between prefusion height carpels and monitored the developmental state of the epidermal cells. In all experiments, regardless of barrier composition or pore size, the epidermal cells redifferentiated and assumed the character of the internal cell type (Verbeke and Walker, 1986). Because the barriers prevented the possibility of surface interactions, we tentatively concluded that diffusible factors were indeed responsible for initiating this distinctive change in the fate of the redifferentiating epidermal cells.

Are Other Epidermal Cells Competent to Redifferentiate?

Are the normally redifferentiating cells somehow different from other epidermal cells? The cells which normally redifferentiate are those on the contacting surfaces, but these are the only cells located in a position to chemically communicate with the cells of the opposing carpel. We performed a series of surgical experiments in which one of the carpels was excised at its base and then rotated 180° on its vertical axis, and then grafted back in place so that the normally redifferentiating cells of one of the carpels were now located at the "back" side while the normally non-contacting and non-redifferentiating cells of that carpel were now in a position to touch the epidermal cells of the opposing carpel. In all of these experiments, the "relocated" cells redifferentiated following contact with epidermal cells of the opposing carpel, while the normally redifferentiating cells which had been rotated to the "outside" position remained epidermal (Siegel and Verbeke, 1989). The results of these experiments indicated that all of the carpel epidermal cells were competent to both send and receive the signal.

Can Redifferentiation Factors be Trapped in Porous Barriers?

Knowing that the epidermal cells which normally were located on the back side of the carpels could redifferentiate if they received the proper signal made it possible to use these cells as a bioassay for the presence or absence of the redifferentiation factors. To do this, we placed porous barriers between prefusion height carpels to "load" these barriers with the redifferentiation factors. The loaded barriers were then placed next to the back surface of one of the carpels. In all of these experiments, the normally non-redifferentiating cells changed their fate when presented with the loaded barriers (Siegel and Verbeke, 1989). These experiments supported our previous conclusions (Verbeke and Walker, 1986) that diffusible factors were indeed responsible for this distinctive response, and showed that we could trap the factors in our porous barriers.

WORK IN PROGRESS

All of our evidence to date suggests that more than one factor is required for redifferentiation (i.e., the carpels may each produce a unique but necessary component). We also have evidence that the redifferentiation factors appear to be interchangeable among genera (at least within the family Apocynaceae). We currently are addressing three major questions utilizing the periwinkle experimental system.

Which Carpel is Which?

Both carpels appear to be necessary for redifferentiation to occur. But, if the carpels are identical and producing exactly the same components and receptors, why do the cells not self-induce? All of our evidence to date suggests that the carpels are not identical but are somehow different, and that each is producing a unique but necessary component for redifferentiation. We are currently trying to determine whether we can predict which carpel is which based on ontogenetic age of individual carpels.

What are the Differences in mRNA Between Developmental Stages?

We have initiated a series of experiments to analyze the carpels at the molecular level. We are in the process of constructing cDNA libraries of carpels at various stages of development, from which we hope to obtain stage-specific genes, some of which should be related to the process of epidermal cell redifferentiation.

What are the Basic Physical and Chemical Characteristics of these Factors?

We hope to purify the exudate and try various solvent partitioning systems as well as ion exchange TLC and/or HPLC. We are also conducting experiments to determine the basic class of compounds to which these redifferentiation factors belong. We are exploiting the fact that we can trap the periwinkle redifferentiation factors in porous polycarbonate membranes by treating "loaded" barriers with RNase; alternatively, we can treat unloaded barriers with RNase and then place these RNase-containing barriers between prefusion carpels to analyze for redifferentiation. We are also considering the possibility that our redifferentiation factors may be small peptides. We would like to perform experiments with proteases, similar to those outlined above with the RNase. The controls for these experiments will be critical. Questions we will have to address include – Will the RNase or protease degrade exogenous protein/RNA? Can exogenous RNase/protease degrade protein/RNA in the polycarbonate barriers? What is the effect in general of these exogenous enzymes on the carpel epidermal cells?

SUMMARY

The overall goal of all of the experiments in our laboratory is to achieve a better understanding of the controls of gene expression which govern plant cell differentiation. Unraveling these complex interactions remains one of the most fascinating problems in modern biology.

ACKNOWLEDGMENTS

I would like to thank Margaret Dietrich, Amy Clore, Louise Latterell, and Rick Heupel, without whose assistance this manuscript could not have been completed. The author's work was supported by grants from the Cellular Biosciences/Developmental Biology Program of the U. S. National Science Foundation.

REFERENCES

Boeke, J.H., 1971, Location of the postgenital fusion in the gynoecium of *Capsella bursa-pastoris* (L.) Med. *Acta Bot. Neerl.* 20:570-576.
Boke, N.H., 1947, Development of the adult shoot apex and floral initiation in *Vinca rosea* L. *Amer. J. Bot.* 34:433-439.
-----., 1948, Development of the perianth in *Vinca rosea* L. *Amer. J. Bot.* 35:413-423.
-----., 1949, Development of the stamens and carpels in *Vinca rosea* L. *Amer. J. Bot.* 36:535-547.
Bruck, D.K., and Walker, D.B., 1985, Cell determination during embryogenesis in *Citrus jambhiri*. I. Ontogeny of the epidermis. *Bot. Gaz.* 146:188-195.

Cusick, F., 1966, On phylogenetic and ontogenetic fusions. *in:* "Trends in plant morphogenesis," E. G. Cutter, ed., Longmans, Green & Co., London.

Hake, S., and Freeling, M., 1986, Analysis of genetic mosaics shows that the extra epidermal cell divisions in *Knotted* mutant maize plants are induced by adjacent mesophyll cells. *Nature* 320:621-623.

Hake, S., Vollbrecht, E., and Freeling, M., 1989 Cloning *Knotted*, the dominant morphological mutant in maize, using Ds2 as a transposon tag. *EMBO J.* 8:15-22.

Knox, R.B., and Clarke, A.E., 1984. Cell recognition in plants, *in*: "Developmental control in animals and plants," C.F. Graham and P.F. Wareing, eds. , Blackwell Scientific Publications, Palo Alto.

Moore, R., 1984, Cellular interactions during the formation of approach grafts in *Sedum telephoides* (Crassulaceae). *Can. J. Bot.* 62:2476-2484.

Nishino, E., 1982, Corolla tube formation in six species of Apocynaceae. *Bot. Mag. Tokyo* 95:1-17.

Siegel, B.A., and Verbeke, J.A., 1989, Diffusible factors essential for epidermal cell redifferentiation in *Catharanthus roseus*. *Science.* 244:580-582.

Sinha, N., and Hake, S., 1990, Mutant characters of *Knotted* maize leaves are determined in the innermost tissue layers. *Develop. Biol.* 141:203-210.

Verbeke, J.A., 1989, Stereological analysis of ultrastructural changes during induced epidermal cell redifferentiation in developing flowers of *Catharanthus roseus* (Apocynaceae). *Amer. J. Bot.* 76:952-957.

-----., 1992, Fusion events during floral morphogenesis. *Annu. Rev. Plant Physiol. Plant Mol. Biol.* 43:583-598.

-----, and Walker, D.B., 1985, Rate of induced cellular dedifferentiation in *Catharanthus roseus*. *Amer. J. Bot.* 72:1314-1317.

-----, and -----., 1986, Morphogenetic factors controlling differentiation and dedifferentiation of epidermal cells in the gynoecium of *Catharanthus roseus*. II. Diffusible morphogens. *Planta* 168:43-49.

Walker, D.B., 1975a, Postgenital fusion in *Catharanthus roseus* (Apocynaceae). I. Light and scanning microscopic study of gynoecial ontogeny. *Amer. J. Bot.* 62:457-467.

-----., 1975b, Postgenital carpel fusion in *Catharanthus roseus* (Apocynaceae). II. Fine structure of the epidermis before fusion. *Protoplasma* 86:29-41.

-----., 1975c, Postgenital carpel fusion in *Catharanthus roseus* (Apocynaceae). III. Fine structure of the epidermis during and after fusion. *Protoplasma* 86:43-63.

-----., 1978, Morphogenetic factors controlling differentiation and dedifferentiation of epidermal cells in the gynoecium of *Catharanthus roseus*. I. The role of pressure and cell confinement. *Planta* 142:181-186.

-----, and Bruck, D.B., 1985, Incompetence of stem epidermal cells to dedifferentiate and graft. *Can. J. Bot.* 63:2129-2132.

STRUCTURE AND EXPRESSION OF STYLE-EXPRESSED AND POLLEN-EXPRESSED COMPONENTS OF GAMETOPHYTIC SELF-INCOMPATIBILITY IN *PETUNIA HYBRIDA*

T.L. Sims,[1,3] J.J. Okuley,[2,4] K.R. Clark,[2,5] and P.D. Collins [1,3]

[1]College of Biological Sciences
[2]Department of Molecular Genetics
Ohio State University
Columbus, OH 43210

INTRODUCTION

Self-incompatibility is a genetic barrier to inbreeding that is based on the ability of the style to discriminate between self and non-self pollen. Historically, this discrimination has been shown to be governed by a single genetic locus known as the S-locus (East and Mangelsdorf, 1925; Bateman, 1955). In *gametophytic* self-incompatibility systems, like those of *Petunia hybrida*, the S-locus recognition phenotype is determined by the haploid genotype of the individual pollen grain. Both self and non-self pollen tubes will germinate and begin to grow through the transmitting tract tissue of the style. If there is a match between the S-allele(s) expressed in the style, and that expressed by a pollen tube, the growth of the (self) pollen tube is inhibited in the upper portion of the style. If there is no match between pollen and style recognition specificities, growth of the pollen tube is not inhibited, and it will grow the length of the style to the ovary, where it can function for fertilization and seed set. A contrasting type of self-incompatibility is *sporophytic* self-incompatibility. Here, the recognition phenotype of the pollen grain is determined by the diploid genotype of the pollen parent. In sporophytic self-incompatibility self pollen grains either fail to germinate at all, or if germination occurs, fail to penetrate through the stigmatic cuticle. In recent years, proteins and genes associated with self-incompatibility alleles have been isolated for both sporophytic systems (such as *Brassica*) and gametophytic systems (such as *Petunia* and *Nicotiana alata*). Current molecular evidence indicates that sporophytic and gametophytic self-incompatibility systems evolved independently and function via different mechanisms (Haring *et al.*, 1990; Nasrallah *et al.*, 1991; Sims, 1992a, 1992b).

---

[3]Present address: Plant Molecular Biology Center, Northern Illinois University, DeKalb, IL 60115

[4]Present Address: Inst. of Biological Chemistry, Washington State University, Pullman, WA 99164

[5]Present Address: Children's Hospital, Columbus OH 43201

RESULTS AND DISCUSSION

We have previously isolated cDNA and genomic clones representing three self-incompatibility alleles of *Petunia hybrida* (Clark et al., 1990). The cloning of the S-allele cDNAs was based on the observation that specific conserved proteins could be associated with S-allele breeding behavior (Mau et al., 1986). We used an oligonucleotide, homologous to the conserved N-terminal S-allele domain, to screen cDNA libraries of mature style RNA from *Petunia hybrida* lines segregating for three S-alleles. These libary screens resulted in the isolation of three putative S-alleles (PS1, PS2, PS3), that have been used in subsequent investigations of gametophytic self-incompatibility.

To investigate the nature of genomic sequences encoding the putative S-alleles, the cDNA clones were used in blot hybridizations of genomic DNAs and were also used to screen genomic libraries constructed from the S_1S_1, S_1S_2 and S_3S_3 lines. The DNA blot hybridization studies indicated that each S-allele was encoded by a single-copy gene. Restriction site mapping of the cloned chromosomal regions showed that a large degree of restriction site polymorphism was present among the three alleles. It has been possible to demonstrate that (within the limits of the number of plants used) the inheritance of these RFLP fragments is correlated with S-allele breeding behavior. In contrast to the patterns seen for the S-allele clones, no restriction site polymorphism was observed when the DNA blots were hybridized with several cloned DNAs (rbcS, Cab, actin, and five random floral-enhanced cDNAs) not related to the S-locus. In each case, restriction fragments of identical size were observed in all three backgrounds (Clark et al., 1990; Clark, 1991). Taken together with data for S-associated genes from other species of the Solanaceae (Anderson et al., 1989; Ai et al., 1990), these blot hybridization data indicate that the cloned sequences represent alleles of the S-locus.

Patterns of S-Locus Gene Expression

As part of our interest in regulation of S-locus gene expression, we have carried out a detailed investigation of spatial and temporal patterns of S-locus expression (Clark et al., 1990; Clark, 1991; Clark and Sims, 1992). RNA blot hybridization showed that all three cDNA clones isolated from *Petunia hybrida* hybridized to an mRNA of approximately 900 nt that accumulated to high levels in styles of S_1S_1, S_1S_2 and S_3S_3 lines. Quantification of mRNA levels indicated that the S1 mRNA accumulated to approximately 0.65% of the mRNA mass in S_1S_1 and S_1S_2 styles. The S2 mRNA was even more abundant, representing about 1.5% of the mRNA mass in S_1S_2 styles. Clark et al. (1990), used a quantitative slot-blot assay to measure the relative level of PS1 mRNA at different stages of floral development. PS1 mRNA accumulated to high levels over the course of floral development, with a large increase between -3 days and maturity. Bud pollination of flowers from the S_1S_1 line at different stages of development demonstrates that mature flowers, and floral buds one or two days prior to anthesis, are capable of preventing fertilization by self-pollen, whereas pollination of floral buds three days prior to anthesis results in capsule formation and seed set. Thus, the greatest accumulation of S1 mRNA is seen during the period when the style undergoes the transition from self-compatibility to self-incompatibility.

We have also isolated non S-locus clones for genes that are preferentially expressed in pistil tissue. The temporal expression pattern of these genes, together with that of the S-locus, defines at least four different patterns of gene expression. Figure 1 shows the relative timing of accumulation of S1 mRNA compared to that for three of these PFE (petunia-floral-enhanced) clones.

Although the S-mRNA is accumulated preferentially in styles, it is not style-specific. Both PS1 and PS2 mRNAs have been observed to accumulate in ovary and petal mRNA populations, although the level of the S-mRNA in these tissues is extremely low (Clark et al., 1990; Clark, 1991; Clark and Sims, 1992). In contrast to the abundant levels seen in styles, S-mRNAs in petals and ovaries are estimated to comprise less than 5×10^{-4} % of the mRNA mass, a level corresponding to about one transcript per cell, averaged over the entire organ (Goldberg et al., 1987). RNase protection assays indicated that the 5' end of the mRNA accumulating in petals was identical to that found in styles, and therefore, that the petal and ovary mRNAs likely were transcribed from the the same gene as is expressed in styles. Accumulation of the S-mRNA sequences could not be detected in leaves, in mature anthers, immature or mature pollen, or in pollen germinated *in vitro* (Clark, 1991; Clark and Sims, 1992). The role of S-allele sequences in ovary and petals is not yet clear.

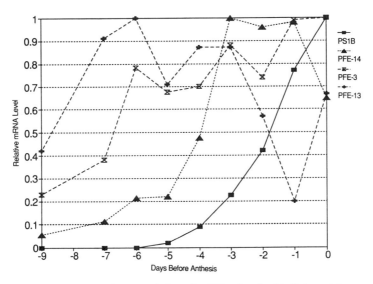

Figure 1. Relative pattern of accumulation of mRNA for the S1 allele and for three non-S-locus clones preferentially expressed in pistils.

Structure and Function of Style-Expressed S-Allele Proteins

Sequence analysis demonstrates that *Petunia* S-alleles have a mixture of conserved and variable domains. Figure 2 shows a similarity plot (Devereaux et al., 1984) for six *Petunia* S-alleles. Relatively conserved regions (C1-C5) are seen as peaks above the line showing the average similarity over the entire protein. Valleys (V1-V4) below that line correspond to regions that are relatively variable in amino acid sequence. Hydropathy plots (Clark et al., 1990, Ioerger et al., 1991) show that domains C1, C4, and C5 are quite hydrophobic. The function of these domains is unkown; they may form the hydrophobic structural core of the protein. An alternative possiblility for the C1 domain is for this region to function in either the binding or import of the S-protein into pollen tubes. The single intron found in these genes interrupts region V2 (Clark, 1991; Ioerger et al., 1991; Okuley, 1991).

The conserved domains C2 and C3 are hydrophilic in character, and have strong homology to active site regions of a widespread class of ribonucleases (Kawata et al., 1988; Horiuchi et al., 1988; McClure et al., 1989; Watanabe et al., 1990; Jost et al., 1991; Ide et al., 1991: Taylor and Green, 1991). Direct analyses of purified S-allele proteins from *Nicotiana alata*, *Petunia hybrida*, and *Petunia inflata* have demonstrated that these proteins have ribonuclease activity *in vitro* (McClure et al., 1989, Broothaerts et al., 1991; Singh et al.,

1991) McClure *et al.* (1990) further demonstrated that pollen tube RNA was degraded *in vivo* during incompatible pollinations, but not during compatible pollinations. Because of the ribonuclease properties associated with the S-allele proteins from *Petunia*, they are now commonly referred to as S-RNase proteins.

A Model for the Mechanism of S-Allele Function

The remarkable pattern of sequence conservation/variability among the different S-alleles and the observation that gametophytic S-allele proteins are ribonucleases, has contributed to a working model that proposes that S-associated ribonuclease activity is an integral component of pollen tube rejection. According to this model, S-RNase proteins are exported and accumulate in the intercellular spaces of the transmitting tract. Self and non-self pollen tubes encounter the S-RNase proteins during growth through the transmitting tissue, with the S-RNase being taken up into incompatible pollen tubes where it acts to degrade pollen tube RNA. Two alternative variations of the model exist. One variation predicts that recognition (mediated by the hypervariable domains) determines whether uptake occurs, with the S-RNase being imported only into incompatible pollen tubes *in vivo*. This model predicts that the pollen component of self-incompatibility would be either a membrane or wall-bound receptor protein. The alternative possibility is that S-RNases are imported into both compatible and incompatible pollen tubes, but that the compatible pollen tubes produce a specific inhibitor of the S-RNase activity.

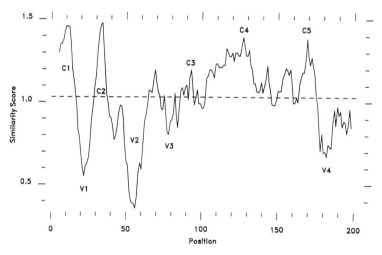

Figure 2. Similarity plot of S-allele protein homologies for alleles isolated from *Petunia* (Clark *et al.*, 1990, Ioerger *et al*, 1990).

Gene Transfer Studies Involving Gametophtyic S-Alleles

To date, although there is a great deal of data showing that the putative S-alleles from *Petunia* (as well as those from *Nicotiana* and *Solanum*) are correlated with self-incompatibility, there has been no direct proof that these genes can confer recognition specificity of pollen tubes. The best experiments to directly demonstrate that the cloned sequences are indeed S-alleles are those utilizing gene transfer to demonstrate co-transfer of an S-allele recognition specificity with the transfer of a defined DNA fragment. In an attempt to test the ability to change breeding specificity via gene transfer, we utilized a strategy where the *Petunia hybrida* S3 allele was transferred into plants homozygous for the S1 allele. Several plants were regenerated from kanamycin selection, assayed for transfer and expression of the S3 gene sequences, and tested for alteration of breeding behavior. To our surprise, many of the plants appeared to have either lost the S3 sequence during regeneration, or showed rearrangement of the transferred DNA, relative to the original construct. Plants mock-transformed with vector alone showed no such problems. In no case did we observe detectable expression of the transferred gene. Although all of the regenerated plants remained self-incompatible, they behaved uniformly as S_1S_1 homozygous lines, and all of the regenerated plants set good seed capsules with S3 pollen.

It is not yet clear why the transferred genes were lost or rearranged in the transgenic plants. Similar results have been obtained by other laboratories investigating gametophytic self-incompatibility (Haring *et al.* 1990, Kaufmann *et al.*, 1991). The failure to obtain expression is not a case of classic co-suppression (Napoli *et al.*, 1990), because expression of the endogenous S-alleles was unaffected.

Transient Expression of S-Locus Sequences

An alternative methodology to stable gene transfer for monitoring gene expression, is to analyze the transient expression of chimeric gene sequences following biolistic particle bombardment of plant cells. To further investigate the potential for expression of the S-RNase sequences in pollen, and to begin to analyze those sequences regulating the specific developmental expression of the S-locus, we used microprojectile bombardment to introduce chimeric S-allele/ß-glucuronidase (GUS) genes into plant cells for transient expression assays (Jefferson *et al.*, 1986, 1987; Klein *et al.*, 1988). When we used a chimeric gene consisting of 1.9 kb of the PS1 promoter sequences fused to GUS, stained cells were observed following particle bombardment of *Petunia* stigmas in areas that correspond to tissue contiguous with the style transmitting tract and demonstrated to express S-alleles *in vivo*. No staining was observed in styles bombarded with a promoter-less GUS plasmid, or with tungsten particles alone. GUS staining was also observed in petals and in epidermal cells of the ovary wall. No histochemical staining was ever observed following bombardment of leaves, anthers, mature pollen or pollen germinated *in vitro*, even though control contructs with CaMV35S or LAT52 promoter regions gave good staining in all of these tissues.

To begin to analyze those sequences required for the correct developmental expression of the S-locus, we constructed deletion clones having variable amounts of S1 5' flanking sequence fused to the ß-glucuronidase reporter gene. Constructs having 8 kb, 1.9 kb, 1.7 kb, 900 bp, 425 bp, 320 bp, 223 bp, and 69 bp of 5' flanking DNA all showed positive histochemical staining in style and petal tissue following bombardment. A construct having only 19 base pairs of 5' flanking DNA, and lacking a TATA box, however, showed no expression by histochemical staining. The histochemical data were largely confirmed using fluorescence assays to give quantitative measurements of GUS activity. The construct having 69 base pairs of 5' flanking sequence, however, gave levels of GUS activity that were only about 40% of the levels observed with longer promoters. This suggests that this region may contain elements important for S-locus gene regulation. The -69 construct does not represent an unregulated "minimal" promoter, as it is not expressed in leaves.

The Pollen-Component of Self-Incompatibility and the Organization of S-Allele Chromosomal Regions

Previous models of self-incompatibility, based on the inability to observe recombination between style and pollen S-locus specificities, proposed that the style and pollen components of the system were identical. The S-allele sequences that have been cloned from *Petunia hybrida* are detectably expressed only in styles, ovaries, and petals. No sequences homologous to the cloned cDNA sequences can be detected in leaves or in pollen at levels of sensitivity that correspond to less than one mRNA molecule per ten cells. Methodology used in attempts to detect pollen expression in gametophytic self-incompatible species has included sensitive RNA blot hybridization, particle gun bombardment assays, and PCR. The inability to detect pollen expression of S-allele sequences implies that either these sequences are encoded by a separate (but tightly linked) transcription unit from that for the style-expressed sequences, or that the pollen-expressed sequences are only expressed following pollination *in vivo*, perhaps in response to signals received from the pistil. However, experiments indicating that purified S-allele proteins can inhibit pollen-tube elongation *in vitro* in an allele-preferential manner would suggest that the pollen-S component has been expressed by the time the pollen is shed (Gray *et al.*, 1991).

To determine if other transcription units were linked to the style-expressed S-alleles, we mapped the location of transcripts in a region of 25 kb of DNA surrounding the *Petunia hybrida* S1 allele. In one approach, labeled cDNAs were transcribed from polyadenylated RNA isolated from different tissues and hybridized to DNA blots of individual genomic clones. This approach identified a sequence region 1.5 kb to 3 kb downstream of the S1 allele that was homologous to an RNA that accumulated in leaves. RNA blot hybridization using cloned genomic fragments indicated that this region encoded a 1.45 kb mRNA. A second sequence region, located approximately 12 kb upstream of the S1 S-RNase coding region showed homology to RNAs that accumulated in petals and ovaries. No hybridization to this region was seen with cDNAs from style, leaf or germinated pollen RNA.

Preliminary DNA blots with labeled cDNA from pollen germinated *in vitro* showed weak hybridization to EcoR1 fragments encoding the S1 S-RNase region. To determine if the faint signals observed actually represented accumulation of RNA in pollen, a 4.2 kb HindIII fragment containing the S1 S-RNase along with 2 kb of 5' flanking DNA and 1 kb of 3' flanking DNA was hybridized to a blot containing polyadenylated RNAs from leaves, petals, ovaries and germinated pollen. This blot showed that a rare-class 2.2 kb transcript from germinated pollen RNA hybridized to the 4.2 kb transcript, along with the 900 bp S-RNase mRNA that accumulates in petals and ovaries. Based on the relative intensity of signal compared to the previously characterized petal and ovary RNAs, this transcript represents about $5\text{-}10 \times 10^{-4}\%$ of the pollen mRNA mass. To further localize the sequences encoding this apparent pollen transcript, the blot was stripped and re-hybridized with a 1.5 kb HindIII-Pst subclone from the 5' region of the 4.2 kb HindIII fragment. This blot also showed hybridization of the 2.2 kb pollen transcript, but hybridization to the petal and ovary RNAs was no longer observed. A similar experiment, using the 3' Pst-HindIII subclone, showed hybridization to the petal and ovary RNAs but not to germinated pollen RNA. Preliminary ribonuclease protection experiments indicate that the direction of transcription of this gene is identical to the S1-RNase. The putative pollen-expressed gene has been tentatively named PAS1. Figure 3 shows a summary of the mapped genes in the region surrounding the S1-RNase sequence.

The S1 stylar-expressed S-RNase shows limited, but detectable, cross homology with the S2 and S3 S-RNase sequences. To determine if sequences homologous to the putative S1 pollen transcript were found linked to other S-alleles, digests of genomic clones spanning the S2 and S3 gene regions were hybridized with a 1.9 kb XbaI fragment predicted to contain major portions of the pollen coding region. Under low stringency hybridization conditions, this fragment hybridized to cloned fragments tightly linked to the S2 and S3 alleles. Unlike the S1 locus, however, the homologous regions linked to the S2 and S3 alleles were 3' to the genes, relative to the direction of S-RNase transcription.

Genetic Studies on Breakdown of Self-Incompatibility

One approach to studying self-incompatibility is to compare the structure and expression of S-alleles in self-incompatible varieties with that in related pseudo-self-compatible varieties showing qualitative or quantitative defects in the self-incompatibility response. Clark *et al.*

(1990) reported that the self-compatible *Petunia hybrida* cultivar MSU1093 contained genomic sequences with homology to the S1 allele. The expression level of the MSU1093 allele was indistinguishable from that of the S1 allele in mature styles. We have recently investigated another line (80-15-5) that arose from selfed progeny of a sib mating of a pollen-part mutant investigated by Dana and Ascher (1986). This line also contains S-allele sequences showing homology by DNA blot hybridization to the S1 allele. Furthermore, the band pattern observed following blot hybridization of genomic DNA digested with different enzymes is identical to that seen for MSU1093, suggesting that the two lines may have the same S-allele. Both MSU1093 and 80-15-5 set large numbers of seed on selfing. Crossing the two lines with MSU1093 as the seed parent gave seed set equal to that obtained by

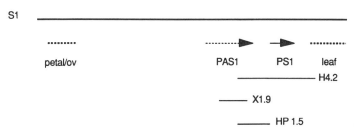

Figure 3. Transcription map of chromosomal region surrounding the S1 allele. Dashed lines labeled petal/ov and leaf represent approximate position of transcripts expressed at low levels in these organs. The arrow labeled PS1 shows the location and direction of transcription of the S1 allele. The arrow labeled PAS1 shows the location and presumed transcription orientation of a region expressed at low levels in germinated pollen. The relative location of DNA fragments used as hybridization probes (see text) is also shown.

selfing either of the lines. The reciprocal cross, with the 80-15-5 line as the seed parent, was self-incompatible. No capsules were formed and no seed set was obtained. These data suggest that the MSU1093 line may be a stylar-inactivating line, and the 80-15-5 line a pollen-inactivating line, both affecting the same allele.

REFERENCES

Ai, Y., Singh, A., Coleman, C.E., Ioerger, T.R., Kheyr-Pour, A., and Kao, T.-H., 1990, Self-incompatibility in *Petunia inflata*: isolation and characterization of cDNAs encoding three S-allele-associated proteins. *Sex. Plant. Reprod.* 3:130.

Anderson, M.A., McFadden, G.I., Bernatsky, R., Atkinson, A. Orpin, T.,Dedman, H., Tregear, G., Fernley, R., and Clarke, A.E., 1989, Sequence variability of three alleles of the self-incompatibility gene of *Nicotiana alata*. *Plant Cell.* 1:483.

Bateman, A.J., 1955, Self-incompatibility systems in angiosperms III. Cruciferae, *Heredity*. 9:53.

Broothaerts, W.J., Vanvinckenroye, P., Decock, B., Van Damme, J., and Vendrig, J.C., 1991, *Petunia hybrida* S-proteins: ribonuclease activity and the role of their glycan side chains in self-incompatibility, *Sex. Plant Repro.* 4:258.

Clark, K.R., Okuley, J.J., Collins, P.D. and Sims, T.L., 1990, Sequence variability and developmental expression of S-alleles in self-incompatible and pseudo-self-compatible petunia. *Plant Cell.* 2:815.

Clark, K.R., 1991, Molecular characterization of S-locus expression in *Petunia hybrida*, Ph.D. Thesis, Department of Molecular Genetics, Ohio State University.

Clark, K.R. and Sims, T.L., 1992, submitted.

Dana, M.N., and Ascher, P.D., 1986, Sexually localized expressin of pseudo-self compatibility (PSC) in *Petunia x hybrida hort*. 1. Pollen inactivation, *Theor. Appl. Genet.* 71:573.

Devereux, J. Haeberli, P. and Smithies, O., 1984, A comprehensive set of sequence analysis programs for the VAX, *Nucl. Acids Res.* 12:387.

East, E.M. and Mangelsdorf, A.J., 1925, A new interpretation of the hereditary behaviour of self-sterile plants, *Proc. Nat. Acad. Sci. U.S.A.* 11:166.

Goldberg, R.B., Hoschek, G., Kamalay, J.C. and Timberlake, W.E., 1978, Sequence complexity of nuclear and polysomal RNA in leaves of the tobacco plant, *Cell.* 14:123.

Gray, J.E., McClure, B.A., Bonig, I., Anderson, M.A., and Clarke, A.E., 1991, Action of the style product of the self-incompatibility gene of *Nicotiana alata* (S-RNase) in *in vitro*-grown pollen tubes, *Plant Cell.* 3:271.

Haring, V., Gray, J.E., McClure, B.A., Anderson, M.A., and Clarke, A.E., 1990, Self-incompatibility: a self-recognition system in plants, *Science.* 250:937.

Horiuchi, H., Yanai, K., Takagi, M., Yano, K., Wakabayashi, E., Sanda, A., Mine, S., Ohgi, K., and Irie, M., 1988, Primary structure of a base non-specific ribonuclease from *Rhizopus niveus*, *J. Biochem.* 103:408.

Ide, H., Kimura, M., Arai, M., and Funatsu, G., 1991, The complete amino acid sequence of ribonuclease from the seeds of bitter gourd (*Momordica charantia*), *FEBS Lett.* 284:161.

Ioerger, T.R., Gohlke, J.R., Xu, B., and Kao, T.-H., 1991, Primary structural features of the self-incompatibility protein in the solanaceae, *Sex. Plant. Reprod.* 4:81.

Jefferson, R.A., Burgess, S.M., and Hirsh, D.A., 1986, β-glucuronidase from Escherichia coli as a gene-fusion marker, *Proc. Nat. Acad. Sci. U.S.A.* 83:8447.

Jefferson, R.A., 1987, Assaying chimeric genes in plants: the GUS gene fusion system, *Plant Molecular Biol. Rep.* 5:387.

Jost, W., Bak, M., Glund, K., Terpstra, P. and Beintema, J.J., 1991, Amino acid sequence of an extracellular phosphate-starvation-induced RNase from cultured tomato (*Lycopersicon esculentum*) cells, *Eur. J. Bioch.* 198:1.

Kaufmann, H., Salamini, F., and Thompson, R.D., 1991, Sequence variability and gene structure at the self-incompatibility locus of *Solanum tuberosum*, *Mol. Gen. Genet.* 226:457.

Kaufmann, H., Kirch, H.-H., Wenner, T., and Thompson, R., 1991, The relationship of major pistil proteins of *Solanum tuberosum* to self-incompatibility, *in*: "Third International Congress of Plant Molecular Biology, Abstracts", R.B. Hallick, ed., University of Arizona, Tucson.

Kawata,Y, Sakiyama, F., and Tamaoki, H., 1988, Amino-acid sequence of ribonuclease T_2 from Aspergillus oryzae, *Eur. J. Biochem.*, 176:683.

Klein, T.M., Gradziel, T., Fromm, M.E. and Sanford, J.C., 1988, Factors influencing gene delivery into *Zea mays* cells by high-velocity microprojectiles, *Bio/technology*. 6:559.

McClure, B.A., Haring, V., Ebert, P.R., Anderson, M.A., Simpson, R.J., Sakiyama, F., and Clarke, A.E., 1989, Style self-incompatibility gene products of *Nicotiana alata* are ribonucleases, *Nature*. 342:955.

McClure, B.A., Gray, J.E., Anderson, M.A. and Clarke, A.E., 1990, Self-incompatibility in *Nicotiana alata* involves degradation of pollen rRNA, *Nature*. 347:757.

McKeon, T.A., Lyman, M.L., and Prestamo, G., 1991, Purification and characterization of two ribonucleases from developing tomato fruit, *Arch. Biochem. Biophys.* 290:303.

Napoli, C., Lemieux, C., and Jorgensen, R., 1990, Introduction of a chimeric chalcone synthase gene into petunia results in reversible co-suppression of homologous genes *in trans*, *Plant Cell*, 2:279.

Nasrallah, J.B., Nishio, T., and Nasrallah, M.E., 1991, The self-incompatibility genes of *Brassica*: expression and use in genetic ablation of floral tissues, *Annu. Rev. Plant Physiol. Plant Mol. Biol.* 42:393.

Okuley, J.J. and Sims, T.L., 1991, A pollen transcript is closely linked to the S-locus in self-incompatible petunia, *Plant Physiol.*, 96S:5.

Okuley, J.J., 1991, Structure, function, and organization of the self-incompatibility locus of *Petunia hybrida*. Ph.D. thesis, Department of Molecular Genetics, Ohio State University.

Sims, T.L., 1992, Genetic regulation of self-incompatibility, *CRC Critical Reviews in Plant Science*, in press.

Sims, T.L., 1992, Molecular genetics of gametophytic self-incompatibility in *Petunia hybrida*, *in* "Genetic Control of Self-Incompatibility and Reproductive Development in Flowering Plants", E.G. Williams, R.B. Knox, and A.E. Clarke, eds., Kluwer Academic Press, Dordrecht, in press.

Singh, A., Ai, Y., and Kao, T.-H., 1991, Characterization of ribonuclease activity of three S-allele associated proteins of *Petunia inflata*. *Plant Physiol.*, 96:61.

Taylor, C.B. and Green, P.J., 1991, Genes with homology to fungal and S-gene RNases are expressed in *Arabidopsis thaliana*, *Plant Physiol.* 96:980.

Watanabe, H., Naitoh, A., Suyama, Y., Inokuchi, N., Shimada, H., Koyama, T., Ohgi, K., and Irie, M., 1990, Primary structure of a base non-specific and adenylic preferential ribonuclease from *Aspergillus saitoi*, *J. Biochem.* 108:303.

LEAFY CONTROLS MERISTEM IDENTITY

IN *ARABIDOPSIS*

Detlef Weigel and Elliot M. Meyerowitz

Division of Biology 156-29
California Institute of Technology
Pasadena, CA 91125
USA

INTRODUCTION

Flower development can be broken down into at least five steps: (1) Upon floral induction, the vegetative shoot meristem is converted into an inflorescence meristem. (2) The inflorescence meristem starts to generate floral meristems, or is itself transformed into a floral meristem. This step can be preceded by the generation of a limited number of secondary inflorescence meristems by the primary inflorescence meristem. (3) The floral meristems produce floral organ primordia. (4) The floral organ primordia adopt different fates according to their position within the developing flower. (5) The floral organ primordia differentiate into floral organs. Despite many efforts, very little is known about the molecules directing these processes. Since classical physiological approaches toward understanding flower development have met only with limited success, a genetic-molecular approach has recently been chosen by several groups (e.g., Komaki et al., 1988; Bowman et al., 1989, 1991, 1992; Hill and Lord, 1989; Kunst et al., 1989; Sommer et al., 1990; Yanofsky et al., 1990; Irish and Sussex, 1990; Carpenter and Coen, 1990; Coen et al., 1990; Martinez-Zapater and Somerville, 1990; Drews et al., 1991; Goto et al., 1991; Koornneef et al., 1991; Schultz and Haughn, 1991; Schultz et al. 1991; Shannon and Meeks-Wagner, 1991; Alvarez et al., 1992; Schwarz-Sommer et al., 1992; Jack et al., 1992; Huijser et al., 1992; Weigel et al., 1992; Huala and Sussex, 1992). The underlying rationale is to first identify mutations that specifically affect different steps of flower development, then to analyze these mutations at the genetic level, and finally to clone the corresponding genes to determine their function at the molecular level. Since the techniques of insertional mutagenesis (e.g., Carpenter and Coen, 1990; Feldmann, 1991) and chromosomal walking (e.g., Meyerowitz, 1989) allow for cloning of a gene based solely on its mutant phenotype, nothing has to be known in advance about the proteins encoded by these genes. Several floral control genes have recently been cloned (Sommer et al., 1990; Yanofsky et al., 1990; Coen et al., 1990; Jack et al., 1992; Huijser et al., 1992; Weigel et al., 1992).

In this review, we will focus on the genetic and molecular characterization of a gene, *LEAFY (LFY)*, that plays a role in the second step of floral development, the production of floral meristems by the inflorescence meristem.

MATURE PHENOTYPE AND DEVELOPMENT OF WILD-TYPE *ARABIDOPSIS* INFLORESCENCES

The aerial part of a mature *Arabidopsis thaliana* plant consists of a rosette of vegetative leaves, and a primary inflorescence shoot bearing a small number of cauline leaves and a potentially indeterminate number of flowers (Fig. 1A). In the axils of the cauline leaves, secondary inflorescence shoots arise, which repeat the pattern of the primary shoot. In contrast to many other dicot species, the flowers of wild-type *Arabidopsis thaliana* are not subtended by small leaves commonly known as bracts (Weberling, 1981). The flowers consist of four concentric rings, or whorls, of organs. The first, outermost whorl contains four sepals, the second whorl four petals, the third whorl four long and two short stamens, and the fourth, innermost whorl two fused carpels that constitute the central gynoecium (Fig. 2A). The leaves and flowers on the inflorescence shoot arise in a helical pattern, whereas the organs in the flower arise in a whorled fashion (Fig. 3A). Furthermore, flowers are determinate, whereas inflorescence shoots are indeterminate.

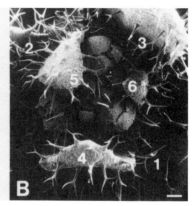

Figure 1. Immature inflorescences of wild-type *Arabidopsis* (A) and a strong *lfy* mutant (B). The wild-type inflorescence bears only two cauline leaves, whereas the mutant inflorescence bears six cauline leaves. Scale bars, 100 μm.

MATURE PHENOTYPE AND DEVELOPMENT OF *LEAFY* MUTANT INFLORESCENCES

The first mutant allele of the *LFY* locus, *lfy-1*, was isolated by Haughn and Somerville (1988) and described in more detail by Schultz and Haughn (1991). Several more alleles have since been isolated (Feldmann, 1991; Schultz and Haughn, 1991; Weigel et al., 1992; Huala and Sussex, 1992; L. Conway and S. Poethig, pers. communication; P. Perez and D. Gerentes, pers. communication). All mutant alleles affect the structure and phenotype of primary and higher-order inflorescence shoots, whereas the vegetative parts of the mutant plants are not altered (Schultz and Haughn, 1991; Weigel et al., 1992; Huala and Sussex, 1992). According to the severity of their phenotype, the *lfy* alleles can be arranged in a phenotypic series (Weigel et al., 1992; Huala and Sussex, 1992). Common to all alleles is an increase in the number of cauline leaves subtending secondary inflorescence shoots (Fig. 1B). The flowers that are eventually produced are abnormal to varying degrees, and exhibit characteristics normally associated with inflorescence shoots. In flowers of weak alleles, sepals and carpels are largely normal, but the number of petals and stamens is reduced, and there are often sepal/petal, petal/stamen, or stamen/carpel mosaic organs. In flowers of intermediate alleles, the number of petals and stamens is further reduced, and almost normal petals and stamens are very rarely found. In the strongest alleles, the flowers consist almost entirely of sepals and carpels and sepal/carpel mosaic organs (Fig. 2B). The sepals have often attributes of leaves such as stellate trichomes and stipules at their base. Most flowers are subtended by bracts whose epidermal morphology is similar to that of cauline leaves (Fig. 2B). Other characteristics of secondary shoots found in *lfy* mutant flowers are occasional internode elongation between leaf-like sepals, and the presence of secondary flowers. Further support for the notion that *lfy* flowers have partial inflorescence shoot character comes from the comparison of early development of *lfy* flowers with that of wild-type flowers. In contrast to wild-type flowers, many organs arise in a helical, as opposed to whorled, fashion in *lfy* mutants (Fig. 3B).

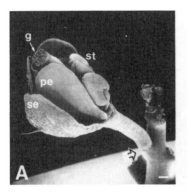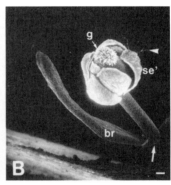

Figure 2. Flowers of wild-type *Arabidopsis* (A) and a strong *lfy* mutant (B). Four different types of floral organs, sepals (se), petals (pe), stamens (st), and carpels forming the central gynoecium (g), can be distinguished in wild-type flowers. In the mutant, only sepal-like (se') and carpel-like organs, which form an abnormal gynoecium, are visible. The sepal-like organs have often stellate trichomes (arrowhead), which are typical for leaves. The mutant flower is subtended by a leaf-like bract (br), which is flanked by stipules (arrow). Bracts are not found subtending wild-type flowers (arrow). Scale bars, 100 µm.

The phenotype of *lfy* mutants can be interpreted in two ways. *lfy* mutations can be said to have a heterochronic effect (cf. Ambros and Horvitz, 1984), because the first phase of inflorescence development, the production of secondary inflorescences, is longer. Although *lfy* mutant inflorescences finally switch to the production of flowers, the flowers are always abnormal, and the switch remains incomplete. A different way to interpret the *lfy* mutant phenotype is as a transformation of flowers into inflorescence shoots, or of floral meristems into inflorescence meristems. The early-arising flowers are completely transformed into shoots, and the later-arising flowers are partially transformed. The interpretation of *lfy* mutations as having a heterochronic effect would emphasize the effect of *lfy* mutations on inflorescence development, whereas the interpretation of *lfy* mutations as affecting floral meristem identity emphasizes the effect of *lfy* mutations on the development of the floral primordia. Since the expression pattern of *LFY* (see below) indicates that *LFY* acts locally within the floral primordia, we think that the primary role of *LFY* is to determine floral meristem identity as opposed to inflorescence meristem identity.

INTERACTION OF *LEAFY* WITH OTHER MUTATIONS AFFECTING FLOWER DEVELOPMENT

Several other mutations affecting flower and inflorescence development in *Arabidopsis* have been described. These mutations fall into three classes. The first class affects, like *lfy*, meristem identity, and includes *apetala1 (ap1)* (Irish and Sussex, 1990), *terminal flower* (Shannon and Meeks-Wagner, 1991; Alvarez et al., 1992), and *cauliflower* (Bowman, 1991). The second class, comprising the cadastral genes, affects whorl identity and/or the expression domains of homeotic genes and includes *superman* (synonym *flo10*; Schultz et al., 1991; Bowman et al., 1992), *apetala2 (ap2)*, and *agamous (ag)*. *ag* and *ap2* belong also to the third class of homeotic mutants that includes two more loci, *apetala3 (ap3)* and *pistillata (pi)* (Bowman et al., 1989, 1991; Hill and Lord, 1989; Kunst et al., 1989; Jack et al., 1992). Since *lfy* mutations do not cause a complete conversion of flowers into inflorescence shoots, the interaction of *lfy* mutants with other meristem identity mutants and with floral homeotic mutants is of particular interest.

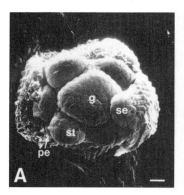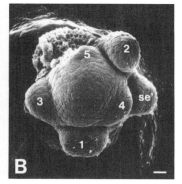

Figure 3. Developing flowers of wild type (A) and a strong *lfy* mutant (B). Some of the outer sepals have been removed to reveal younger organ primordia. Organs arise in a whorled manner in wild type, but in a helical fashion in the mutant. The primordia in the mutant flower are numbered according to the temporal sequence in which they apparently arose. Scale bars, 10 µm.

The homeotic genes themselves fall into three classes. Each class is required for determining organ identity in two adjacent whorls of organs. In addition, *ag* and *ap2* mutations affect organ number in the flower. *ag* mutant flowers are indeterminate and have thus many more floral organs than wild type, whereas the number of organs in strong *ap2* mutant flowers is severely reduced. This effect is apparently mediated by ectopic expression of the *AG* wild-type product in *ap2* mutants (Bowman et al., 1991). Two of the four wild-type organ types, petals and stamens, are almost completely missing in strong *lfy* alleles. In *ap3* and *pi* mutants, petals and stamens are homeotically transformed into sepals and carpels, respectively. As expected, *ap3; lfy* and *pi; lfy* double mutants exhibit essentially the *lfy* single mutant phenotype (Weigel et al., 1992; Huala and Sussex, 1992), suggesting that the activity of the *AP3* and *PI* wild-type products is severely reduced or absent in strong *lfy* mutants. In situ hybridization to tissue sections of *lfy* mutants shows that the RNA levels of *AP3*, which has been cloned by Jack et al. (1992), are indeed severely reduced in *lfy* mutants (D.W., T. Jack, E.M.M., unpublished results).

Carpels are homeotically transformed into sepals in *ag* mutants. Interestingly, carpels do develop in *ag; lfy* double mutants (Weigel et al., 1992; Huala and Sussex, 1992). This result can be interpreted in two ways: Either, in *lfy* mutants, carpel development does not require *AG* activity. Or, in *ag* mutants, *LFY* activity interferes with the development of carpels. Currently, we cannot distinguish between these two alternatives. In contrast to *lfy* flowers, *ag* flowers are indeterminate, and this trait is found in *ag; lfy* double mutants, indicating that this aspect of *AG* function is active in *lfy* mutants.

Finally, sepals are homeotically transformed into leaves and carpels in *ap2* mutants. In *ap2-1; lfy* as well as in *ap2-2; lfy* double mutants, no sepals are observed (Weigel et al., 1992; Huala and Sussex, 1992), indicating that the organ identity function of *AP2* is active in *lfy* mutants.

In contrast to organ identity, *lfy* is epistatic to both *ap2* and *ag* with respect to the pattern of organ emergence, which is largely spiral in strong *lfy* mutants, but whorled in *ap2* and *ag* mutants (Weigel et al., 1992).

Why do *lfy* mutants fail to transform all flowers completely into shoots, as indicated by the presence of floral organs and by the residual activity of floral homeotic genes in *lfy* mutants? The molecular analysis of *LFY* suggests that the strongest alleles represent a complete loss of function (Weigel et al., 1992), making it unlikely that residual *LFY* activity is the cause for the remaining floral characteristics in strong *lfy* mutants. Another possibility is that *LFY* is not the only factor determining floral meristem identity, but that *LFY* interacts with one or more other factors in this process. Two other mutations have been reported to affect floral meristem identity, *ap1-1* and a weak mutant allele of the *AP2* locus, *ap2-1*. In both mutants, first-whorl sepals are transformed into cauline leaf-like bracts, and secondary flowers arise in the axils of the first whorl organs, although the presence of a leaf-like organ is not a prerequisite for the emergence of secondary flowers, and vice versa (Bowman et al., 1989; Irish and Sussex, 1990). *ap1-1; ap2-1* double mutants show a synergistic effect (Irish et al., 1990). The exact role of *AP2* in determining floral meristem identity remains to be elucidated, since strong *ap2* mutants do not produce secondary flowers, and the formation of secondary flowers in *ap1-1* is largely suppressed by the strong *ap2-2* allele (Bowman, 1991; Bowman et al., 1991).

Double mutants between *ap1-1* and both weak and strong *lfy* alleles show an enhancement of the *lfy* mutant phenotype (Weigel et al., 1992; Huala and Sussex, 1992). Mostly leaf-like organs arise in a spiral fashion, although some of the later organs still exhibit carpelloid characteristics. A similar phenotype is found in a double mutant with *ap2-1* (Huala and Sussex, 1992; D.W. and E.M.M., unpublished data). Huala and Sussex (1992) reported that carpelloid organs are absent in *ap2-1; lfy-1* and *ap2-1; lfy-13* double mutants. *ap2-1; lfy-6* double mutants, however, still produce carpelloid organs (D.W. and E.M.M., unpublished data). *lfy-1* and *lfy-6* are associated with the same molecular lesion in the *LFY* coding region (Weigel et al., 1992), eliminating the possibility that these differences are due to different severity of the *lfy* mutations. Genetic background, however, might play a role, because *lfy-6* is in the *er* background, whereas *lfy-1* and *lfy-13* are in the *Er*$^+$ background. Again, the exact role of *AP2* in this process remains enigmatic, since an *ap2-2; lfy-6* double mutant does not show a stronger conversion of floral meristems into inflorescence meristems than *lfy-6* alone, although there are more leaf-like organs in this double mutant than in *lfy-6* single mutants (Weigel et al., 1992).

MOLECULAR ANALYSIS OF THE *LEAFY* LOCUS

The genomic region containing the *LFY* gene was isolated by chromosomal walking, and the gene has subsequently be localized within the cloned DNA by virtue of its homology to the *Antirrhinum* gene *FLORICAULA (FLO)*, which had been previously cloned by Coen et al. (1990). Ten mutant alleles have single codon changes in this gene, identifying it as the

LFY gene (Weigel et al., 1992; D.W. and E.M.M., unpublished). In addition, an eleventh allele is associated with a rearrangement of the *LFY* transcribed region (P. Perez and D. Gerentes, unpublished; cited in Weigel et al. [1992]). The protein sequences of *LFY* and *FLO* are not similar to other proteins of known biochemical function, although it has been speculated that *FLO* and *LFY* might be transcription factors (Coen et al., 1990).

As a first step toward understanding the molecular basis of *LFY* function, its expression in developing flowers was analyzed by in situ hybridization. *LFY* is expressed in floral primordia, before any floral organ primordia start to emerge (Fig. 4). This finding is consistent with the notion that *LFY* controls floral meristem identity, and with the finding that *lfy* mutations affect both the identity and the pattern of organs arising on the floral meristem. The earliest expression of *LFY* can be detected in the anlagen of the floral primordia, i.e., before the floral primordia become morphologically visible as buttresses on the flanks of the inflorescence meristem. This is consistent with the analysis of early development of *lfy* mutant flowers, since the earliest effects of *lfy* mutations are observed as soon as floral buds would form on the wild-type inflorescence apex (Weigel et al., 1992; Huala and Sussex, 1992). In *lfy* mutants, bracts develop in place of the floral buds, and mutant floral buds form only later in the axils of the bracts.

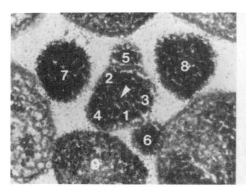 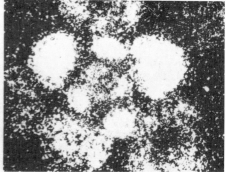

Figure 4. Expression of *LFY* RNA as revealed by in situ hybridization to a transverse section of an *Arabidopsis* inflorescence apex. Left – phase contrast view, right – dark field view. *LFY* RNA is absent from the center of the apex, the inflorescence meristem proper (arrowhead), but is detected in the floral primordia produced at the flanks of the inflorescence apex. Numbers indicate the apparent order, in which the floral primordia arose, with '1' being the youngest. '1' is an anlage of a floral primordium, '2' to '4' are stage 1 primordia, '5' and '6' are stage 2 primordia, '7' is a stage 3 primordium, and '8' and '9' are stage 4/5 primordia. Stages are according to Smyth et al. (1990).

LFY is expressed uniformly throughout the floral primordium until stage 3, when the first sign of morphological differentiation, the emergence of sepal primordia, appears. Interestingly, uniform *LFY* expression is detected until after the region-specific expression of the homeotic genes *AG*, *AP3*, and *PI* is initiated (Drews et al., 1991; Jack et al., 1992; K. Goto and E.M.M., unpublished). This makes it unlikely that *LFY* provides direct positional cues for the region-specific expression of the homeotic genes. Consistent with this observation, we have found that a strong *lfy* mutation affects mainly the RNA level rather than the domain of expression of the *AP3* gene (D.W., T. Jack, and E.M.M., unpublished).

LFY is expressed at later stages in petals, in the filaments of the stamens, and in carpels (Weigel et al., 1992). In contrast to the early expression pattern, the pattern of late *LFY* expression is different from the one of *FLO*, which is not expressed in stamens (Coen et al., 1990). What the function of the late expression of *LFY* is, remains to be elucidated. To address this problem, we are currently introducing transgenes that carry the *LFY* coding region under the control of an inducible promoter into *lfy* mutants. With these transgenes, it should be possible to separate early and late *LFY* function.

CONCLUSIONS

Flower development is a process that can be subdivided into a series of steps. Genetic analysis has shown that different classes of mutations specifically affect different steps of flower development. Only few mutations affect more than one step.

A pivotal step in flower development is the establishment of the floral primordium, or the generation of a floral meristem by the inflorescence meristem. A main factor controlling this step is the product of the *LFY* gene, mutations in which cause a partial transformation of flowers into inflorescence shoots (Schultz and Haughn, 1991; Weigel et al., 1992; Huala and Sussex, 1992). *LFY* appears to act in concert with other factors, since even complete loss-of-function mutations do not effect a complete conversion of floral into inflorescence meristems (Weigel et al., 1992). One of these additional factors is the *AP1* gene, and circumstantial genetic evidence indicates that another factor is encoded by, or regulated by, the *AP2* gene (Weigel et al., 1992; Huala and Sussex, 1992; J. Bowman, Z. Liu, and E.M.M., unpublished).

Molecular cloning of the *LFY* gene has revealed that it is expressed before the homeotic genes analyzed to date, namely *AG*, *AP3* and *PI* (Weigel et al., 1992). Further analysis is required to understand how *LFY* interacts at the molecular level with other meristem identity genes and with the homeotic genes.

ACKNOWLEDGMENTS

We thank E. Huala and I. M. Sussex for a preprint of their *LFY* manuscript. D.W. was supported by an EMBO Long-term Fellowship, and by a Senior Fellowship from the American Cancer Society, California Division. Our work on *LEAFY* is supported by US Department of Energy Division of Energy Biosciences grant DE-FG03-88ER13873 to E.M.M.

REFERENCES

Alvarez, J., Guli, C. L., Yu, X.-H., and Smyth, D. R. (1992) *terminal flower*: A gene affecting inflorescence development in *Arabidopsis thaliana*. Plant J. *2*, 103-116.

Ambros, V., and Horvitz, H. R. (1984) Heterochronic mutants of the nematode *Caenorhabditis elegans*. Science *226*, 409-416.

Bowman, J. L. (1991) Molecular genetics of flower development in *Arabidopsis thaliana*. Ph. D. thesis, California Institute of Technology.

Bowman, J. L., Smyth, D. R., and Meyerowitz, E. M. (1989) Genes directing flower development in *Arabidopsis*. Plant Cell *1*, 37-52.

Bowman, J. L., Smyth, D. R., and Meyerowitz, E. M. (1991) Genetic interactions among floral homeotic genes of *Arabidopsis*. Development *112*, 1-20.

Bowman, J. L., Sakai, H., Jack, T., Weigel, D., Mayer, U., and Meyerowitz, E. M. (1992) *SUPERMAN*, a regulator of floral homeotic genes in *Arabidopsis*. Development *114*, 599-615.

Carpenter, R., and Coen, E. S. (1990) Floral homeotic mutations produced by transposon-mutagenesis n *Antirrhinum majus*. Genes Dev. *4*, 1483-1493.

Coen, E. S., Romero, J. M., Doyle, S., Elliott, R., Murphy, G., and Carpenter, R. (1990) *floricaula*: a homeotic gene required for flower development in Antirrhinum majus. Cell *63*, 1311-1322.

Drews, G. N., Bowman, J. L., and Meyerowitz, E. M. (1991) Negative regulation of the *Arabidopsis* homeotic gene *AGAMOUS* by the *APETALA2* product. Cell *65*, 991-1002.

Feldmann, K. A. (1991) T-DNA insertion mutagenesis in *Arabidopsis*: mutational spectrum. Plant J. *1*, 71-82.

Goto, K., Kumagai, T., and Koornneef, M. (1991) Flowering responses to light-breaks in photomorphogenic mutants of *Arabidopsis thaliana*, a long-day plant. Physiol. Plant. *83*, 209-215.

Haughn, G. W., and Somerville, C. R. (1988) Genetic control of morphogenesis in *Arabidopsis*. Dev. Genet. *9*, 73-89.

Hill, J. P., and Lord, E. M. (1989) Floral development in *Arabidopsis thaliana*: comparison of the wildtype and the homeotic *pistillata* mutant. Can. J. Bot. *67*, 2922-2936.

Huala, E., and Sussex, I. M. (1992) *LEAFY* interacts with floral homeotic genes to regulate *Arabidopsis* floral development. Plant Cell, *submitted*.

Huijser, P., Klein, J., Lönnig, W.-E., Meijer, H., Saedler, H., and Sommer, H. (1992) Bracteomania, an inflorescence anomaly, is caused by the loss of function of the MADS-box gene *squamosa* in *Antirrhinum majus*. EMBO J. *11*, 1239-1249.

Irish, V. F., and Sussex, I. M. (1990) Function of the *apetala-1* gene during *Arabidopsis* floral development. Plant Cell *2*, 741-751.

Jack, T., Brockman, L. L., and Meyerowitz, E. M. (1992) The homeotic gene *APETALA3* of Arabidopsis thaliana encodes a MADS-box and is expressed in petals and stamens. Cell *68*, 683-697.

Komaki, M. K., Okada, K., Nishino, E., and Shimura, Y. (1988) Isolation and characterization of novel mutants of *Arabidopsis thaliana* defective in flower development. Development *104*, 195-203.

Koornneef, M., Hanhart, C. J., and Vanderveen, J. H. (1991) A genetic and physiological analysis of late flowering mutants in *Arabidopsis thaliana*. Mol. Gen. Genet. *229*, 57-66.

Kunst, L., Klenz, J. E., Martinez-Zapater, J., and Haughn, G. W. (1989) *AP2* gene determines the identity of perianth organs in flowers of *Arabidopsis thaliana*. Plant Cell *1*, 1195-1208.

Martinez-Zapater, J. M., and Somerville, C. R. (1990) Effect of light quality and vernalization on late-flowering mutants of *Arabidopsis thaliana*. Plant Physiol. *92*, 770-776.

Meyerowitz, E. M. (1989) *Arabidopsis*, a useful weed. Cell *56*, 263-269.

Schultz, E. A., and Haughn, G. W. (1991) *LEAFY*, a homeotic gene that regulates inflorescence development in Arabidopsis. Plant Cell *3*, 771-781.

Schultz, E. A., Pickett, F. B., and Haughn, G. W. (1991) The *FLO10* gene product regulates the expression domain of homeotic genes *AP3* and *PI* in Arabidopsis flowers. The Plant Cell *3*, 1221-1237.

Schwarz-Sommer, Z., Hue, I., Huijser, P., Flor, P. J., Hansen, R., Tetens, F., Lönnig, W.-E., Saedler, H., and Sommer, H. (1992) Characterization of the *Antirrhinum* floral homeotic MADS-box gene *deficiens*: evidence for DNA binding and autoregulation of its persistent expression throughout flower development. EMBO J. *11*, 251-263.

Shannon, S., and Meeks-Wagner, D. R. (1991) A mutation in the Arabidopsis *TFL1* gene affects inflorescence meristem development. Plant Cell *3*, 877-892.

Smyth, D. R., Bowman, J. L., and Meyerowitz, E. M. (1990) Early flower development in *Arabidopsis*. Plant Cell *2*, 755-767.

Sommer, H., Beltrán, J. P., Huijser, P., Pape, H., Lönnig, W.-E., Saedler, H., and Schwarz-Sommer, Z. (1990) *Deficiens*, a homeotic gene involved in the control of flower morphogenesis in *Antirrhinum majus*: the protein shows homology to transcription factors. EMBO J. *9*, 605-613.

Weberling, F. (1981) Morphologie der Blüten und der Blütenstände. (Stuttgart: Eugen Ulmer Verlag).

Weigel, D., Alvarez, J., Smyth, D. R., Yanofsky, M. F., and Meyerowitz, E. M., (1992) *LEAFY* controls floral meristem identity in *Arabidopsis*. Cell *69*, in press.

Yanofsky, M. F., Ma, H., Bowman, J. L., Drews, G. N., Feldmann, K. A., and Meyerowitz, E. M. (1990) The protein encoded by the *Arabidopsis* homeotic gene *agamous* resembles transcription factors. Nature *346*, 35-39.

ARABIDOPSIS AS A MODEL SYSTEM FOR ANALYSIS OF LEAF SENESCENCE AND INFLORESCENCE-MERISTEM LONGEVITY

Linda L. Hensel and Anthony B. Bleecker

University of Wisconsin
Department of Botany
Madison, WI 53706

ABSTRACT

The longevity of the Arabidopsis plant is a function of both the life span of the somatic tissue and the extent to which the meristems produce new somatic tissues. We have chosen to study the leaf as a model for somatic-tissue senescence, and the primary inflorescence meristem as a model for meristem longevity. Under constant light, temperature and humidity, we found that the individual adult rosette leaves and the primary inflorescence meristem have a determined life span. Arabidopsis' life strategy is monocarpic, meaning the plant dies in association with the cessation of reproduction. We asked if senescence ensued as a result of reproduction by determining if we could uncouple reproduction from senescence. By analyzing single-gene mutants either delayed or defective in reproduction, we found no relationship between rosette-leaf life span and reproduction. In contrast, we did find that primary inflorescence-meristem longevity is nearly doubled in a male-sterile line where reproduction does not occur. We are using both individual leaf life spans and primary meristem longevity as markers to determine the role of heritable traits in the regulation of senescence and whole plant longevity. We are comparing leaf senescence and meristem proliferative ability in different ecotypes, existing developmental and hormonal mutants, and newly isolated mutants.

INTRODUCTION

We are interested in determining the genetic components that play a role in defining the life span of the Arabidopsis plant. Two basic processes govern the longevity of the plant: 1) the ability of the plant to produce new vegetative tissues through the continued activity of the meristems, and 2) the longevity of those somatic tissues once they have been produced. Arabidopsis is well-suited for these studies, because it is well defined developmentally, has a short life cycle, and can easily be grown under controlled conditions. For example, under constant growing conditions the primary inflorescence meristem produces 30 to 40 flowers before undergoing apical arrest about 50 days post-germination (Shannon and Meeks-Wagner, 1991; Alvarez et al., 1992). In addition, the longevity of somatic tissues is easily followed, because their development is well documented (Medford et al., 1992) and somatic tissues once produced, senesce progressively in the same order in which they emerge from the meristem. To simplify our analyses, we have chosen the primary inflorescence as a model for meristem longevity and adult rosette leaves as a model for somatic tissue senescence. We are defining "meristem longevity" as proliferative capacity, in contrast to "somatic tissue longevity" which measures the actual life span of specific organ systems from initiation to senescence. The process of somatic senescence is easily monitored in photosynthetic tissues based on the loss of chlorophyll followed by total tissue collapse. In this paper we focus on somatic tissue senescence and discuss the utility of Arabidopsis as a model system for the identification of genetic factors which influence tissue longevity.

The final developmental stage of most somatic tissues in plants is characterized by an organized series of processes collectively referred to as the senescence syndrome (reviewed in Nooden, 1988; Kelly and Davies, 1988). Initial stages include macromolecular turnover and loss of plastid function. Nucleic acids, proteins, and lipids are broken down into transportable molecules, for example amino acids, which can then be delivered to the developing reproductive organs (Thimann, 1980). The chloroplast loses assimilatory capacity in the early stages and in the final stages is itself broken down for salvage. It is thought that a complex set of salvage pathways organize the breakdown of the cell components. These processes are thought to be controlled by genes specifically activated during the syndrome (Woolhouse, 1983). The regulation of these genes has not been elucidated. However, a myriad of environmental factors can induce the senescence syndrome. For example, extreme temperatures, excessive water or drought, nutrient deprivation, and pathogen invasion will induce senescence, resulting in a shorter life span for the tissue (Levitt, 1980a and 1980b; Sprague, 1964; Butler and Simon, 1971). In addition to stress-induced senescence, the signal(s) of post-reproductive development that may induce senescence in monocarpic plants are hypothesized, but their identity remains elusive.

Several hypotheses have been developed to explain the onset of the senescence syndrome associated with reproduction (reviewed in Nooden, 1988; Kelly and Davies, 1988). In monocarpic plants such as Arabidopsis, where the plant dies in association with reproduction, correlative control hypotheses have been developed. This seems biologically sensible because, if the reproductive tissue in some way controlled the onset of somatic tissue senescence, the developing reproductive organs would be guaranteed nutrients at the appropriate time. One hypothesis states that a hormone or chemical messenger that initiates the syndrome is sent from the reproductive tissues to the somatic tissues to induce senescence (McCollum, 1934; Murneek, 1951). A second hypothesis proposes that the reproductive organs' nutrient drain on the plant vegetative tissues influences the timing of senescence (Molisch, 1928). A third hypothesis, which does not involve correlative controls, states that the photosynthetic tissues simply "age" and as a result senesce. The aging hypothesis has been dismissed by many who study senescence (Nooden, 1988; Kelly and Davies, 1988), because the timing of senescence can be experimentally altered (see below).

One major advantage to using Arabidopsis in senescence studies is that hormone and developmental mutants are available to address proposed hypotheses of senescence. In addition, there are several dozen ecotypes (geographic races) available for QTL (quantitative trait loci) analysis as was done for other characteristics in different plant species (Tanksley et al., 1982; Edwards et al., 1992; Diers et al., 1992). Arabidopsis also has a short generation time, a well-characterized genetic map, and requires relatively little growing space (Meyerowitz, 1989); thus, isolation of new mutants is feasible. Studies with both existing and new mutants may allow identification of the heritable traits that contribute to the senescence syndrome in Arabidopsis.

We used available mutants in Arabidopsis to address the proposed correlative control hypotheses that explain the onset of somatic tissue senescence in monocarpic plants. The true test of the correlative control hypothesis would be to determine if reproduction could be uncoupled from the senescence syndrome. We used available lines with single-gene mutations which either delayed flowering time or caused sterility and measured both somatic tissue and meristem longevity to address this question. We further tested the nutrient drain theory by analysis of a developmental mutant with an altered source: sink ratio. In Arabidopsis, the terminal flower mutant, has wild type vegetative growth, but the inflorescence meristems terminate after only one to three flowers are produced, reducing the "sink" more than ten fold (Alvarez et al., 1992; Shannon and Meeks-Wagner, 1991). We, thus, measured the longevity of somatic tissues in the terminal flower mutant. In addition to testing the primary hypothesis of correlative controls, mutants are available to test for fine-tuned regulation of the senescence syndrome.

Plant hormones have been implicated in being involved in influencing the timing of senescence (reviewed in Nooden, 1988; Kelly and Davies, 1988). Prior work has shown the senescence syndrome to be either induced or delayed by different hormonal applications.

For example, ethylene-treated leaves lose chlorophyll, RNA, and protein as they do during senescence (Aharoni and Lieberman, 1979; Kao and Yang, 1983; Abeles, 1973; Lieberman, 1979). In contrast to ethylene-induced senescence, cytokinin has been shown to delay senescence in detached rice leaves (Kao and Yang, 1983), soybean explants (Neumann et al., 1983), and sunflower leaves (Kende, 1964 and 1965). In Arabidopsis, we have the opportunity to examine the role, if any, a particular hormone has in the senescence syndrome because a battery of different hormone mutants have been isolated.

ARABIDOPSIS DEVELOPMENT

Arabidopsis grows vegetatively as a basal rosette for several weeks, and then produces a primary inflorescence which puts forth secondary branches (Figure 1) (Müeller, 1961). The number of leaves produced in the basal rosette varies with flowering time (Martinez-Zapater and Somerville, 1990) and varies among ecotypes. For example, Wassilewskija (WS) strains typically have a basal rosette of 5 to 8 leaves, while the Columbia (Col) strain has 8 to 12 rosette leaves. The first two leaves to emerge are defined as juvenile, the second pair of leaves as intermediate, and the remaining rosette leaves as adult (Medford et al., 1992). The primary stalk typically produces one to three aerial secondary branches, each subtended by a cauline leaf. Subsequent to secondary branches, 30 to 50 flowers, which result in siliques (fruits) when fertilized, are formed on the primary stalk. In addition, several secondary and tertiary branches, each subtended by a cauline leaf, develop and produce flowers and fruits.

Figure 1. A diagram of a four week old Arabidopsis plant. The basal rosette leaves and primary-stalk aerial-cauline (C) leaves are numbered by order of emergence.

EXPERIMENTAL DESIGN

In order to study age-related senescence that minimizes environmental stresses, we have chosen to grow plants in a constant environment. The system we use for Arabidopsis is described in Figure 2.

We have selected rosette leaves one through six and the first two aerial cauline leaves of the primary stalk (Figure 1) for initial studies of age-related senescence in a somatic tissue. The life span for each individual leaf is measured as the time of visual emergence (< 1mm) to the time of visual senescence or degreening.

Meristem longevity is measured as the total number of flowers produced by the primary meristem in the plant's life time.

ROSETTE-LEAF LIFE SPANS

When Arabidopsis is grown in constant light, temperature, humidity, and with a continuous supply of a water-nutrient mixture, the senescence syndrome occurs in an

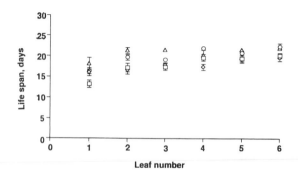

Figure 2. A 20 x 15 inch flat has four 1/2 inch wicks (Hummert Seed 18-4043, St. Louis) running the interior length of the flat. The flat is suspended above a second flat with a continuous supply of 10% Hoaglands (1938) into which wicks are inserted. The plants are grown in a 2:1 mixture of Jiffy Mix: perlite (Jiffy Products of America, Batavia, Illinois):on top of a 1/4-1/2 inch layer of perlite. Plants are evenly spaced to that rosette leaves do not overlap (approximately 50 per flat depending on ecotype). Plants were grown under a mixture of fluorescent and incandescent (100-150 $\mu E\ m^{-2}\ s^{-1}$) at 22°C, with 65-80% humidity in a Conviron growth chamber (Asheville, North Carolina).

individual leaf at a defined time post leaf-emergence. Figure 3 shows individual-leaf life span measurements of the Col strain for four separate experiments. The juvenile leaves have a slightly shorter life span than the intermediate and adult leaves. In addition, the juvenile and intermediate-leaf life spans encompass a three to five day range among different experiments. However, the adult-leaf life spans among the four experiments differ by only one or two days. Thus, with growing parameters described in Figure 2, reproducible life span measurements for adult leaves are made.

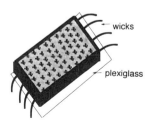

Figure 3. The leaf life span was measured as the time of visual emergence (<1 mm in length) to the time of ≥ 50% degreening. Leaves are numbered by their order of emergence, as depicted in Figure 1. Leaves one and two emerged concomitantly; thus, leaf one was scored as the leaf that visually senesced first. The mean life span ± SE is shown (SE < 0.5 are not shown). Exp. 1 ( ), n=29; Exp. 2 (◊), n=10; Exp. 3 (O), n=16; Exp. 4 (Δ), n=6.

Two parameters, ≥ 50% degreening and total tissue collapse, were used to define life span. The results for Col and WS strains are shown in Figure 4A and 4B. The visual assay of chlorophyll loss (≥ 50% degreening) consistently occurs two to three days prior to total tissue collapse in these strains. Thus ≥ 50% degreening, which is easily visible, is an appropriate way to measure life span. We have used this method to compare life spans of several ecotypes (geographic races).

The results of one life-span comparison between ecotypes Col and WS, are shown in Figure 4C. The life spans between the Col and WS strains studied are not dramatically different; however, the WS leaves do live two to three days longer. Aerial-cauline leaf senescence was also measured in the same manner as rosette leaf senescence (Figure 4C, columns 7 and 8) and is indistinguishable from the adult rosette-leaf life spans for the ecotypes examined. Initial studies of additional ecotypes show the adult rosette-leaf life span of other ecotypes may differ by as much as ten days (data not shown).

We have begun to address the correlative control hypotheses discussed in the introduction: 1) the nutrient drain hypothesis, and 2) the hormonal or chemical messenger hypothesis, and their relationship to rosette-leaf senescence. The key experiments would be designed to determine if reproduction can be uncoupled from somatic tissue senescence. We have measured rosette-leaf life span for lines with single-gene mutations resulting in a delayed-flowering time, sterility, and reproductive-mass reduction and have surprisingly found no affect on leaf life span (manuscript in preparation). This is in direct contrast to the hypothesis that longevity is a function of the time of reproduction. In addition, we do not observe any alteration in senescence in hormonal mutants, such as gibberellic-acid (GA) deficient and abscisic-acid (ABA) insensitive mutants. Since many hormonal mutants still produce or react to small amounts of the hormone, our initial studies cannot rule out the involvement of GA or ABA. The data does indicate, however, that rosette-leaf senescence occurs independent of a reproductive signal or a nutrient drain from reproductive organs. The tissue has a defined life span and seems to senesce simply as a result of "aging".

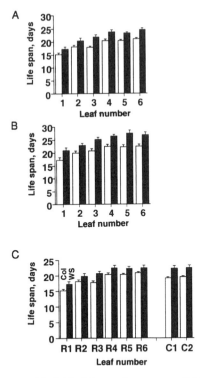

Figure 4. Life spans of individual leaves for ecotypes Col (**A**) and WS (**B**). Life spans were measured as the time from visual leaf emergence (< 1 mm) to ≥ 50% degreening (open bars) or to the time of total tissue collapse (solid bars). Rosette leaves 1 through 6, numbered by order of emergence are shown. (**C**) Col (open bars) and WS (closed bars) life spans from leaf emergence to ≥ 50% degreening for rosette leaves (R1 through R6) and the first two aerial cauline leaves on the primary stalk (C1 and C2) are compared. The mean life span ± SE is shown. Col R1 through R6, n=62; Col C1 and C2, n=29; WS, n=24.

PRIMARY MERISTEM LONGEVITY

Whole plant senescence is governed by both the life span of somatic tissues, such as the leaf, and the ability of the meristems to proliferate. As stated, we measured meristem longevity as the total number of flowers produced. Meristem longevity can also be studied in available single-gene mutants, ecotypes, and new mutants. In contrast to rosette-leaf senescence, we found that fertilization does play a role in inflorescence-meristem longevity (manuscript in preparation). A male-sterile line produced nearly twice the number of flowers as the wild-type plant on the primary meristem. Fertilization of the male-sterile flowers at floral emergence reduced the proliferative capacity of the primary meristem approximately 20%. Thus, some post-fertilization event signals the meristem to cease proliferation. The hypotheses discussed for the senescence syndrome of somatic tissue may have some bearing on inflorescence meristem longevity. The signal could be a "messenger" sent from the reproductive organs to the apical meristem, or, pod fill, itself, may adsorb nutrients or growth factors that the meristem needs to sustain longevity.

Even though male-sterile inflorescence meristems have a greater proliferative capacity than fertilized inflorescence meristems, they still eventually cease proliferation. Thus, by decreasing reproduction, meristem longevity can only be extended. The coupling between reproduction and longevity has been shown for several animal species, as well (reviewed in Finch, 1991).

DISCUSSION

The Arabidopsis plant is not designed to live forever. Interactions between somatic and meristematic tissue govern the life span of an Arabidopsis plant. Molisch first hypothesized that selection would act on monocarpic plants, as in animals, where "the shortest life span consistent with a normal reproductive period" is favored (1928). By applying Molisch's hypothesis to Arabidopsis, one could predict a short life span for individual rosette leaves, as well as the whole plant. However, the fact that the rosette-leaf life span is uncoupled from reproduction was surprising. The rosette leaf life plan has evolved in such a way that a reproduction-associated signal to senesce is unnecessary for the species survival.

In contrast to rosette-leaf senescence, we found that primary-inflorescence meristem longevity is influenced by the reproductive status of the plant. In the absence of fruit development, the primary-inflorescence meristem produces nearly twice the number of flowers. Thus, the most efficient body plan will encompass 1) somatic tissues that are maintained the minimal time required to get the reproductive tissues established, and 2) meristematic tissues of the inflorescence that are under correlative control providing insurance for reproductive success.

In order to understand components that contribute to plant longevity, we have chosen to study both somatic tissue senescence and meristem proliferative capacity in *Arabidopsis thaliana*. Both existing and new mutants that have leaves or inflorescence meristems with altered longevity could allow identification of developmental events or signals involved in processes that determine plant longevity. In addition, long-lived and short-lived ecotypes will be useful in determining 1) the number of gene(s) involved in longevity by quantitative trait analysis and 2) the types of gene products that confer longevity. Future efforts will be directed toward defining parameters that influence the complex pathways between and within somatic and meristematic tissues that define plant longevity.

REFERENCES

Abeles, F.B., 1973, "Ethylene in Plant Biology," Academic Press, New York.
Aharoni, N., and Lieberman, M., 1979, Ethylene as a regulator of senescence in tobacco leaf discs, *Plant Physiol.* 64:801.
Alvarez, J. Guli, C.L., Yu, X. and Smyth, D.R., 1992, *terminal flower*: a gene affecting inflorescence development in *Arabidopsis thaliana*, *Plant Journal* 2:103.
Butler, R.D., and Simon, E.W., 1971, Ultrastructural aspects of senescence in plants, *Adv. Gerontol. Res.* 3:73.

Diers, B.W., Keim, P., Fehr, W.R., and Shoemaker, R.C., 1992, RFLP analysis of soybean seed protein and oil content, *Theor. Appl. Genet.* 83:608.

Edwards, M.D., Helentijaris, T., Wright, S., and Stuber, C.W., 1992, Molecular-marker-facilitated investigations of quantitative trait loci in maize. 4. analysis based on genome saturation with isozyme and restriction fragment length polymorphism markers, *Theor. Appl. Genet.* 83:765.

Finch, C.E., 1990, "Longevity, Senescence, and the Genome," University of Chicago Press, Chicago.

Hoagland, D.R. and Arnon, D.I., 1938, The water-culture method for growing plants without soil, *Calif. Agr. Expt. Sta. Cir.* 347, Berkeley, California.

Kao, C.H.., and Yang, S.F., 1983, Role of ethylene in the senescence of detached rice leaves, *Plant Physiol.* 73:881.

Kelly, M.O. and Davies, P.J., 1988, The control of whole plant senescence, *CRC Critical Reviews in Plant Sciences* 7:139.

Kende, H., 1964, Preservation of chlorophyll in leaf sections by substances obtained from root exudate, *Science*, 145:1066.

Kende, H., 1965, Kinetinlike factors in root exudate of sunflower, *Proc. Natl. Acad. Sci.* USA 53:1302.

Levitt, J., 1980a, "Responses of Plants to Environmental Stresses, Vol. 1, Chilling, Freezing, and High Temperature Stresses," Academic Press, New York.

Levitt, J., 1980b, "Response of Plants to Environmental Stresses, Vol. 2, Water, Radiation, Salt, and Other Stresses," Academic Press, New York.

Martinez-Zapater, J.M., and Somerville, C.R., 1990, Effect of light quality and vernalization on late-flowering mutants of *Arabidopsis thaliana, Plant Physiol.* 92:770.

McCollum, J.P., 1934, Vegetative and reproductive responses associated with fruit development in cucumber, *Mem. Cornell Agric. Exp. Sta.* 163:3.

Medford, J.I., Behringer, F.J., Callos, J.D., and Feldman, K.A., 1992, Normal and abnormal development in the Arabidopsis vegetative shoot apex, *The Plant Cell* 4:631.

Meyerowitz, E.M., 1989, Arabidopsis, a useful weed, *Cell* 56:263.

Molisch, H., 1928, Der lebensdauer der pflanze, translated by F.H. Fulling, 1938, *in* "The Longevity of Plants," H. Fulling, New York.

Müller, A., 1961, Zur charakterisierung der bluten und infloreszenzen von *Arabidopsis thaliana* (L.) Heynh., *Kulturpflanze* 9:364.

Murneek, A.E., 1951, Growth regulators during fertilization and post-fertilization period, *Palest. J. Bot. Hortic.* Sci. 8:8.

Neumann, P.M., Tucker, A.T., and Nooden, L.D., 1983, Characterization of leaf senescence and pod development in soybean explants, *Plant Physiol.* 72:182.

Nooden, L.D., 1988, The phenomena of senescence and aging, *in* "Senescence and Aging in Plants," L.D. Nooden and A.C. Leopold, eds., Academic Press, San Diego.

Shannon, S. and Meeks-Wagner, D.R., 1991, A mutation in the Arabidopsis *TFL1* gene affects inflorescence meristem development, *Plant Cell* 3:877.

Sprague, H.B., ed., 1964, "Hunger Signs in Crops," 3rd Ed., McKay, New York.

Tanksley, S.D., Medina-Filho, H., and Rick, C.M., 1982, Use of naturally-occurring enzyme variation to detect and map genes controlling quantitative traits in an interspecific backcross of tomato. *Heredity* 49:11.

Thimann, K.V., 1980, The senescence of leaves, in "Senescence in Plants," K.V. Thimann, ed., CRC Press Boca Raton, Florida.

Woolhouse, H.W., 1983, The general biology of plant senescence and the role of nucleic acid and protein turnover in the control of senescence processes which are genetically programmed, in "Post-harvest Physiology and Crop Preservation," M. Lieberman, ed., Plenum Press, New York.

ANALYSIS OF A RECEPTOR-LIKE PROTEIN KINASE OF *ARABIDOPSIS THALIANA*

G. Eric Schaller, Sara Patterson, and Anthony B. Bleecker

University of Wisconsin
Department of Botany
Madison, WI 53706

ABSTRACT

Genomic and cDNA clones encoding a novel receptor-like protein kinase (TMK1) have been isolated from *Arabidopsis thaliana*. The predicted protein has an intracellular kinase domain most related to the receptor tyrosine kinases but containing diagnostic serine/threonine sequences, a transmembrane domain, and an extracellular domain containing 11 copies of a leucine rich repeat. Leucine rich repeats have only been found in proteins associated with the plasma membrane and are involved in protein/protein interactions. Domain-specific antibodies against the extracellular and intracellular domains of TMK1 have been made using fusion proteins expressed in *E. coli*. In extracts of *Arabidopsis*, the antibodies specifically immunodecorate a polypeptide of about 120 kD. The native TMK1 protein from *Arabidopsis* is capable of reversible binding to lectin columns, and digestion with endoglycosidase F reduces the apparent molecular mass of the immunodecorated protein by 10 kD, indicating that the native protein is glycosylated. The intracellular domain of TMK1 was expressed as a fusion protein with maltose binding protein in *E. coli* and was found capable of autophosphorylation on serine and threonine residues, indicating that the intracellular domain of TMK1 is a functional protein kinase.

INTRODUCTION

Receptor protein kinases play a central role in the cell-to-cell signaling of animal systems, and mediate the effects of growth factors and hormones such as epidermal growth factor and insulin. These receptor protein kinases have a number of defining characteristics (Yarden and Ullrich, 1988; Ullrich and Schlessinger, 1990). They typically contain an extracellular glycosylated domain which binds the ligand, a single transmembrane domain, and a cytoplasmic domain with tyrosine kinase activity. To date, all known ligands for the receptor protein kinases are polypeptides. The binding of ligand to the extracellular domain of the receptor activates the cytoplasmic kinase domain which autophosphorylates and also phosphorylates substrates within the cell. In this manner an extracellular signal is transduced across the plasma membrane leading to changes within the cell. Of particular interest in animals has been the discovery that many oncogene products are derived from receptor protein kinases. This serves as a measure of the importance of the receptor protein kinases to normal cell growth and can provide an understanding as to their function.

The plasma membrane of higher plants is also a site of hormone action and signal transduction. However, very little is known as to the mechanisms by which plants transduce signals. Some evidence suggests that binding sites for plant growth regulators such as auxin (Barbier-Brygoo et al., 1989), gibberellin (Hooley et al., 1991), and fusicoccin (de Boer et al., 1989) exist in the plasma membrane. A diffusable peptide has also been identified in tomato capable of inducing the synthesis of proteinase inhibitors. This same peptide stimulates kinase-mediated phosphorylation of specific proteins in isolated plasma membranes (Farmer et al., 1989).

Recently, genes have been cloned from higher plants which code for proteins structurally similar to the animal receptor kinases, suggesting that plants might have an analogous mechanism for signal transduction. The first report that higher plants might contain receptor protein kinases similar to those in animals occurred as recently as 1990, following the identification by Walker and Zhang (1990) of a cDNA clone encoding a transmembrane protein kinase in corn. This protein kinase contains an extracellular domain related to S-locus glycoproteins of *Brassica*, a single transmembrane domain, and a cytoplasmic domain with sequences diagnostic of a serine/threonine protein kinase. Since then similar receptor-like protein kinases, all with aminoterminal domains related to the S-locus glycoproteins, have been identified in *Brassica* (Stein et al., 1991) and *Arabidopsis* (Tobias et al., 1992). Taken together these results are indicative of a signaling system in higher plants analogous to that mediated by receptor protein kinases in animal cells.

THE TMK1 GENE PRODUCT OF *ARABIDOPSIS*

Both genomic and cDNA clones encoding a novel protein kinase have recently been isolated from *Arabidopsis thaliana* (Chang et al., 1992). The identified gene (referred to as TMK1) codes for a transmembrane protein kinase having the characteristics one would expect for a receptor protein kinase, but also including several unique features. The major features of the TMK1 polypeptide are illustrated in Figure 1.

The predicted intracellular domain of the TMK1 gene product has homology to protein kinases. Protein kinases have been found to have eleven subdomains of sequence homology common to all kinases sequenced to date (Hanks and Quinn, 1991). The carboxyterminal half of TMK1 has all eleven of these subdomains with the correct spacing and organization. This catalytic kinase domain is most homologous to the catalytic domains of the receptor-like protein kinases previously identified in higher plants, with 39% identity to maize ZmPK1 (Walker and Zhang, 1990) and 35% identity to *Brassica* SRK (Stein et al., 1991). Of particular interest is the observation that TMK1, along with the other plant receptor-like protein kinases, appears to code for a serine/threonine specific kinase, not a tyrosine-specific kinase, as based upon analysis of sequences diagnostic for the specificity of phosphorylation (Hanks and Quinn, 1991). The animal receptor protein kinases are typically tyrosine-specific, with only a few exceptions.

Amino-terminal to the kinase domain of TMK1 is a hydrophobic domain spanning 24 amino acids, characteristic of a membrane-spanning domain (Engelman et al., 1986). The hydrophobic domain is followed by a cluster of basic amino acids indicative of the cytoplasmic side of a transmembrane region (Weinstein et al., 1982). This would place the kinase domain within the cell as expected.

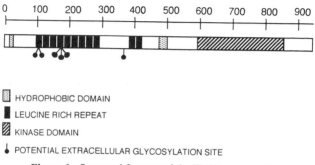

Figure 1. Structural features of the TMK1 gene product.

```
LRR1      LRNLSELERLELQWNNISGPVP
LRR2      LSGLASLQVLMLSNNNFDSIPS
LRR3      FQGLTSLQSVEIDNNPFKSWEI
LRR4      LRNASALQNFSANSANVSGSLP
LRR5      PDEFPGLSILHLAFNNLEGELP
LRR6      SLAGSQVQSLWLNGQKLTGDIT
LRR7      LQNMTGLKEVWLHSNKFSGPLP
LRR8      FSGLKELESLSLRDNSFTGPVP
LRR9      LLSLESLKVVNLTNNHLQGPVP
LRR10     FGAIKSLQRIILGINNLTGMIP
LRR11     LTTLPNLKTLDVSSNKLFGKVP

Consensus L**L**L**L*L**N*α*G*αP
```

Figure 2. Alignment of leucine rich repeats of the TMK1 gene product. The derived consensus sequence is also shown. Hydrophobic amino acids are indicated by α. Nonconserved amino acids are indicated by an asterisk.

The proposed extracellular domain of TMK1 contains 11 imperfect repeats of a 22 amino-acid consensus sequence (Figure 2) arranged in two blocks, one block with nine repeats and the other block with two repeats. This consensus sequence has the features of a leucine rich repeat unit found in a number of other proteins. To date, leucine rich repeats (LRR) have only been found in proteins associated with the plasma membrane, and in all cases they are implicated in protein/protein interactions (Rothberg et al., 1990). Several of these proteins serve as cell-surface receptors, although only the human proto-oncogene *trk* actually encodes a receptor protein kinase (Schneider and Schweiger, 1991). No protein with leucine rich repeats has previously been identified or characterized in higher plants.

In addition to the leucine rich repeats, the extracellular domain of TMK1 has a short hydrophobic sequence at the amino-terminus with the characteristics of a signal sequence. A signal sequence is consistent with the expected plasma membrane localization of a receptor protein kinase. The extracellular domain also has 6 potential glycosylation sites. Glycosylation is typical of the extracellular domains of receptor kinases (Yarden and Ullrich, 1988).

It should be emphasized that the extracellular domain of the TMK1 polypeptide is completely different from the S-protein homologous domains of the receptor-like protein kinases previously identified in higher plants. As such, two separate classes of receptor-like protein kinases have now been identified, indicating the existence of multiple signal transduction pathways mediated by these receptor-like protein kinases.

ANALYSIS OF THE NATIVE GENE PRODUCT OF TMK1 IN *ARABIDOPSIS*

Although genes for receptor-like kinases have been identified in higher plants, little is known about how these proteins actually function in the plant. A method that has proven particularly useful in our examination of this type of question has been the creation of gene fusions, by which we have expressed portions of TMK1 in *E. coli*.

Portions of the LRR domain, the kinase domain, and the carboxyterminal domain were each expressed as fusion proteins with glutathione S-transferase (GST) using the expression vector pGEX-2T, and used to prepare domain-specific antibodies (Figure 3). These antibodies provide a convenient means to examine the native form of the TMK1 polypeptide in *Arabidopsis*. Using the antibodies directed against either the LRR domain or the kinase

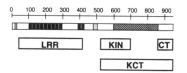
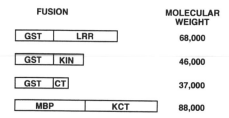

Figure 3. Fusion proteins from the TMK1 gene product. Fusion proteins were made with glutathione S-transferase (GST) to the following domains of the TMK1 gene product: the leucine rich repeat domain (LRR), a portion of the kinase domain (KIN), and the carboxyterminal region (CT). For functional expression, a fusion protein was made with maltose binding protein (MBP) to the entire carboxyterminal half of the TMK1 gene product (KCT).

domain, a single polypeptide of 120 kD was detected in extracts of *Arabidopsis*. This polypeptide was detected in total leaf extracts and in the membrane fraction. A variety of tissue types were probed with the polyclonal antibody to the LRR domain, and the TMK1 polypeptide was found in all tissue types examined, including root, stem, leaf, and floral tissues. As such, it would appear that the TMK1 polypeptide plays a general role in the plant.

All the known animal receptor protein kinases are glycosylated and, taking advantage of their carbohydrate moieties, can be partially purified on lectin columns. *Arabidopsis* membranes were solubilized with 1% Triton X-100 and incubated with various immobilized lectins. The immunodecorated TMK1 polypeptide bound to concanavalin A-agarose which is typical of most plant glycoproteins containing mannose in the oligosaccharide chain (Faye et al., 1989). The TMK1 polypeptide could also bind and be specifically eluted from wheat germ lectin-agarose which binds N-acetylglucosamine residues. Furthermore, we found that treatment of the denatured TMK1 polypeptide from *Arabidopsis* with endoglycosidase-F reduced the apparent molecular mass of the immunodecorated polypeptide by 10,000 daltons. These results indicate that the native TMK1 polypeptide in *Arabidopsis* is a glycoprotein, consistent with its possible role as a receptor protein kinase.

FUNCTIONAL EXPRESSION OF THE TMK1 KINASE DOMAIN IN *E. coli*

The majority of eukaryotic receptor protein kinases examined to date phosphorylate tyrosine residues; however, the TMK1 polypeptide contains sequences typically diagnostic for protein kinases with a specificity for serine and threonine. In order to determine the actual specificity of the TMK1 kinase, the carboxyterminal half of the TMK1 polypeptide was expressed as a fusion protein with the maltose binding protein (MBP) in *E. coli* (Figure 3) using the expression vector pMAL-c. Upon induction of *E. coli* carrying the fusion construct, a polypeptide of the expected molecular weight was produced that was recognized by the polyclonal antibodies directed against the kinase and carboxyterminal domains of the TMK1 polypeptide. Taking advantage of the maltose binding domain, the fusion protein produced was rapidly purified away from other *E. coli* proteins by binding to amylose resin, and then assayed for kinase activity as illustrated in Figure 4. The fusion protein was observed to autophosphorylate, and analysis of the phosphorylated residues indicated the presence of phosphoserine and phosphothreonine but not of phosphotyrosine. These results

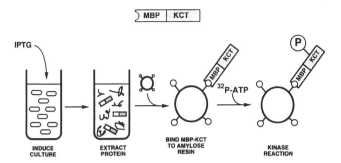

Figure 4. Purification and assay of a fusion protein containing the functional kinase domain of the TMK1 gene product. Expression of the fusion protein (MBP-KCT) was induced in a culture of *E. coli*, total protein extracted, and the fusion protein specifically purified by binding to amylose resin. A kinase reaction was performed directly on fusion protein bound to the resin, and initiated by the addition of γ^{32}P-ATP.

indicate that the TMK1 polypeptide contains a functional kinase domain, and that it is capable of phosphorylating serine and threonine residues.

THE TMK1 GENE PRODUCT AS A CANONICAL RECEPTOR-LIKE PROTEIN KINASE

The polypeptide product of the TMK1 gene has all the canonical features expected in a receptor protein kinase. It also has several intriguing features rare among the receptor protein kinases, in particular the presence of leucine rich repeats and a specificity for serine and threonine residues. Leucine rich repeats have been found in only one other receptor protein kinase, the human proto-oncogene *trk*, in which case three short repeats of the leucine rich repeat core sequence were found (Schneider and Schweiger, 1991). The receptor protein kinases have typically been thought of as tyrosine-specific with respect to phosphorylation, however several receptor protein kinases have recently been identified containing sequences indicating serine/threonine specificity; this specificity has been confirmed for the transforming growth factor-β type II receptor (Lin et al. 1992).

The TMK1 receptor-like protein kinase was originally identified based on coding sequence (Chang et al., 1992). In order to extend the analysis from gene to protein, we have taken advantage of the expression systems developed in *E. coli*. Fusion proteins containing various portions of the TMK1 gene were expressed and purified, and polyclonal antibodies generated from these capable of recognizing specific domains of the TMK1 polypeptide. In this manner, the TMK1 polypeptide was identified in *Arabidopsis* and found to be associated with membranes and present in all tissue types examined. Furthermore, the immunodecorated TMK1 polypeptide was found to be a glycoprotein in its native form. By expressing the entire kinase domain of TMK1 in *E. coli*, we have also been able to analyze the activity of the TMK1 polypeptide, and have found that it represents a serine/threonine protein kinase capable of autophosphorylation.

The biochemical results, like the structural features determined from the coding sequence, are all consistent with the TMK1 polypeptide functioning as a receptor protein kinase in *Arabidopsis*. Based on our knowledge of the receptor protein kinases from animals, further predictions can be made and tested as to characteristics for the plant receptor-like protein kinases. (1) Upon extracellular binding of ligand, the animal receptor kinases dimerize, which leads to activation of the cytoplasmic kinase domains (Ullrich and Schlessinger, 1990). A similar mechanism for activation may apply to the plant receptor-like protein kinases. The leucine rich repeats, which are implicated in protein-protein interactions, of the TMK1 polypeptide could help mediate such a dimerization. (2) Autophosphorylation plays important and varied roles in the animal receptor kinases, where it can lead to ligand-independent activation of the kinase domain (Rosen et al., 1983), removal of a competitive substrate site for the kinase (Honegger et al., 1988), or regulation of interactions with other proteins (Kazlauskas and Cooper, 1989). We have determined that the TMK1 gene encodes a kinase capable of autophosphorylation, but the significance of this autophosphorylation is as yet unknown. (3) Various deletion mutants of animal receptor kinases can have significant effects upon activity and functioning of the kinase. In particular, deletion of the extracellular domain can result in constitutive activation of the kinase, potentially generating an oncogene (Ullrich and Schlessinger, 1990). On the other hand, deletion of the cytoplasmic kinase domain can create a dominant inhibitor of activity in the cell, apparently by interfering with proper dimerization after ligand binding (Ueno et al., 1991). Similar mutants can be constructed from the plant receptor-like protein kinases and their effects studied in the whole plant after transformation.

THE ROLE OF PLANT RECEPTOR-LIKE PROTEIN KINASES IN PLANTS

The question remains as to what roles the receptor-like protein kinases play in plant signal transduction. This question applies just as well to the S-receptor kinases as to the LRR-receptor kinase. Transcripts of the SRK gene, the S-locus receptor kinase of *Brassica*, have been found exclusively in male and female reproductive organs consistent with the proposed role in the self-incompatibility response (Stein et al., 1991). However, very similar S-related receptor kinases have also been found expressed in vegetative tissues and in plant species that do not show self-incompatibility (Walter and Zhang, 1990; Tobias et al., 1992). This indicates that S-receptor kinases, like the LRR-receptor kinase, probably serve a more general role in the plant, aside from a specific involvement in interactions between pollen and pistil.

The presence of the plant cell wall, forming a spatial barrier between cells, prevents the plasma membranes of separate cells from coming in direct contact. As such, extracellular domains of plant receptor-like kinases would interact either with components located at the plasma membrane of their particular cell or with substances capable of diffusing between cells. The later would be consistent with our knowledge of animal receptor kinases which bind diffusable polypeptide ligands (Ullrich and Schlessinger, 1990). In this context, the recent identification of an 18 amino-acid polypeptide capable of inducing defensive genes in plants upon wounding is particularly attractive as a potential ligand (Pearce et al., 1991). Another possibility could be a secondary interaction, where a signaling ligand bound to a diffusable binding protein would interact with the receptor-like kinase; auxin and the auxin-binding protein, which have been reported to be present in the extracellular space, could be modeled as an example of this type of interaction (Crass, 1991). Finally it may be that, unlike the animal receptor kinases, the plant receptor-like kinases are capable of interacting with ligands other than polypeptides: oligosaccharides for example (Farmer et al., 1989).

In animals, the receptor protein kinases are typically involved in transducing signals involved in cell growth and differentiation (Ullrich and Schlessinger, 1990). Plant tissue, however, is composed primarily of permanent tissues no longer capable of cell division. The meristematic tissues, still embryonic and capable of cell division, are localized to specific parts of the plant. The meristematic tissue could potentially be a site where the plant receptor-like protein kinases play a role in growth and differentiation. The plant receptor-like kinases could also play a role in the timing and initiation of specific events. This is already known for the S-locus receptor kinase of *Brassica* and its involvement in self-incompatibility (Stein et al., 1991); another possible role could be in the initiation of flowering. Localization of some of the S- and LRR-receptor transcripts and gene products indicates their presence throughout much of the plant, not just in meristematic or specific tissue types (Walker and Zhang, 1990; Tobias et al., 1991; Chang et al., 1992), suggesting a more general role in the plant. One possibility is that these plant receptor-like kinases could participate in plant defensive responses to wounding and pathogen attack. Self-recognition also appears to be a general characteristic of plants, not limited to the self-incompatibility response, grafting being an example of this.

REFERENCES

Barbier-Brygoo, H., Ephritikhine, G., Klambt, D., Ghislain, M., and Guern, J., 1989, Functional evidence for an auxin receptor at the plasmalemma of tobacco mesophyll protoplasts, *Proc. Natl. Acad. Sci. USA* 86:891.

Chang, C., Schaller, G.E., Patterson, S.E., Kwok, S.F., Meyerowitz, E.M., and Bleecker, A.B., 1992, The TMK1 gene from Arabidopsis codes for a protein with structural and biochemical characteristics of a receptor protein kinase, *Plant Cell* (in press).

Crass, J.W., 1991, Cycling of auxin-binding protein through the plant cell: pathways in auxin signal transduction, *New Biologist* 3:813.

de Boer, A.H., Watson, B.A., and Cleland, R.E., 1989, Purification and identification of the fusicoccin binding protein from oat root plasma membrane, *Plant Physiol.* 89:250.

Engelman, D.M., Steitz, T.A., and Goldman, A., 1986, Identifying nonpolar transbilayer helices in amino acid sequences of membrane proteins. *Ann. Rev. Biophys. Chem.* 15:321.

Farmer, E.E., Pearce, G., and Ryan, C.A., 1989, *In vitro* phosphorylation of plant plasma membrane proteins in response to the proteinase inhibitor inducing factor, *Proc. Natl. Acad. Sci. USA* 86:1539.

Faye, L., Johnson, K.D., Sturm, A., and Chrispeels, M.J., 1989, Structure, biosynthesis, and function of asparagine-linked glycans on plant glycoproteins, *Physiol. Plant.* 75:309.

Hanks, S.K., and Quinn, A.M., 1991, Protein kinase catalytic domain sequence database: Identification of conserved features of primary structure and classification of family members, *Methods Enzymol.* 200:38.

Honegger, A., Dull, T.J., Szapary, D., Komoriya, A., Kris, R., Ullrich, A., and Schlessinger, 1988, Kinetic parameters of the protein tyrosine kinase activity of EGF-receptor mutants with individually altered autophosphorylation sites, *EMBO J.* 7:3053.

Hooley, R., Beale, M.H., and Smith, S.J., 1991, Gibberillin perception at the plasma membrane of *Avena fatua* aleurone protoplasts, *Planta* 183:274.

Kazlauskas, A., and Cooper, J.A., 1989, Autophosphorylation of the PDGF receptor in the kinase insert region regulates interactions with cell proteins. *Cell* 58:1121.

Lin, H.Y, Wang, X.-F., Ng-Eaton, E., Weinberg, R.A., and Lodish, H.F., 1992, Expression cloning of the TGF-β type II receptor, a functional transmembrane serine/threonine kinase, *Cell* 68:775.

Pearce, G., Strydom, D., Johnson, S., Ryan, C.A., 1991, A polypeptide from tomato leaves induces wound-inducible proteinase inhibitor proteins, *Science* 253:895.

Rosen, O.M., Herrera, R., Olowe, Y., Petruzzelli, L.M., and Cobb, M.H., 1983, Phosphorylation activates the insulin receptor tyrosine protein kinase, *Proc. Natl. Acad. Sci. USA* 80:3237.

Rothberg, J.M., Jacobs, J.R., Goodman, C.S., and Artavanis-Tsakonas, S., 1990, slit: An extracellular protein necessary for development of midline glia and commissural axon pathways contains both EGF and LRR domains, *Genes Dev.* 4:2169.

Schneider, R., and Schweiger, M., 1991, A novel modular mosaic of cell adhesion motifs in the extracellular domains of the neurogenic *trk* and *trkB* tyrosine kinase receptors, *Oncogene* 6:1807.

Stein, J.C., Howlett, B., Boyes, D.C., Nasrallah, M.E., and Nasrallah, J.B., 1991, Molecular cloning of a putative receptor protein kinase gene encoded at the self-incompatibility locus of *Brassica oleracea*, *Proc. Natl. Acad. Sci. USA* 88:8816.

Tobias, C.M., Howlett, B., and Nasrallah, J.B., 1992, An *Arabdopsis thaliana* gene with sequence similarity to the S-locus receptor kinase of *Brassica oleracea*-Sequence and expression, *Plant Physiol.* 99:284.

Ueno, H., Colbert, H., Escobedo, J.A., and Williams, L.T., 1991, Inhibition of PDGF β receptor signal transduction by coexpression of a truncated receptor, *Science* 252:844.

Ullrich, A., and Schlessinger, J., 1990, Signal transduction by receptors with tyrosine kinase activity, *Cell* 61:203.

Walker, J.C., and Zhang, R., 1990, Relationship of a putative receptor protein kinase from maize to the S-locus glycoproteins of *Brassica*, *Nature* 345:743.

Weinstein, J.N., Blumenthal, R., van Renswoude, J., Kempf, C., and Klausner, R.D., 1982, Charge clusters and the orientation of membrane proteins, *J. Membr. Biol.* 66:203.

Yarden, Y., and Ullrich, A., 1988, Growth factor receptor tyrosine kinases, *Annu. Rev. Biochem.* 57:443.

ABSTRACTS OF TALKS

Ethylene: An Unique Plant Signalling Molecule
Athanasios Theologis. USDA/Plant Gene Expression
Center-U.C. Berkeley, 800 Buchanan Street, Albany, CA
94710

Ethylene is considered to be the fruit ripening hormone.
Because of its effect on plant senescence, large amounts of fruit
and vegetables are lost annually in the United States and in third
world countries. As a first step towards inhibiting fruit
senescence, we have expressed antisense RNA to ACC synthase in
tomato fruits using the 35S CaMV promoter (1-6). The results
indicate that the antisense RNA inhibits ethylene production, and
more importantly, ripening and softening of the fruits. The
inhibition can be reversed by treatment with exogenous ethylene.
These results also indicate that ethylene is the trigger and not
the by-product of the ripening process. Furthermore, the prospect
arises that inhibition of ethylene production by antisense ACC
synthase RNA can be a general method for preventing senescence in
a variety of fruits and vegetables.

CELL COMMUNICATION AND THE COORDINATION OF
DIFFERENTIATION. J. A. Verbeke. Department of Plant Sciences, University of
Arizona, Tucson, AZ 85721, USA

The overall goal of our work is to understand the mechanisms which underlie
differentiation events in multicellular plants. Postgenital carpel fusion in
Catharanthus roseus, which involves the rapid redifferentiation of four hundred
epidermal cells into parenchyma, has proven to be a useful experimental system for
these studies. Our work to date has shown that: (1) diffusible factors in *C. roseus*
carpels trigger epidermal cell redifferentiation; (2) both abaxial and adaxial carpel
epidermal cells can respond to the stimulus; and (3) each carpel may produce a
unique but necessary component. We are continuing our characterization of the
communication factors that convey positional information to plant epidermal cells by
testing whether developmental age can be correlated with differences in
redifferentiation factors produced. We are also testing the effects of various enzyme
treatments on the redifferentiation factors. We hope to achieve a better
understanding of the controls of gene expression which govern plant cell
differentiation, for we believe that unraveling these complex interactions remains
one of the most fascinating problems in modern biology. The fusing carpels in
Catharanthus roseus provide a particularly useful system with which to tackle these
exciting questions. [Supported by the National Science Foundation.]

GENETIC AND MOLECULAR ANALYSIS OF LIGHT-REGULATED SEEDLING DEVELOPMENT IN *ARABIDOPSIS*. Joanne Chory. Plant Biology Laboratory. The Salk Institute, P.O. Box 85800, San Diego, 92186-5800, USA.

Dicotyledonous seedling development is dramatically affected by light. The transition from heterotrophic to photoautrophic growth, with the associated differentiation of leaves and chloroplasts is dependent on the presence of light and developmental signals. Several regulatory photoreceptors are involved in the perception of light signals, but little is known of the transduction pathways that mediate light-regulated development. We are taking a combined genetic and molecular biological approach to identify potential components of the light-signal transduction pathways. We have identified a class of *Arabidopsis thaliana* mutants that show many characteristics of light-grown plants even when grown in complete darkness. The mutants define 4 complementation groups, designated *det*1, *det*2, *det*3, *det*4 (de-*et*iolated). Because these mutations are both pleiotropic and recessive, we have hypothesized that *DET* genes play a negative regulatory role in photoregulated leaf and chloroplast development in *Arabidopsis*. In a separate set of experiments, work in our lab and by Dr. M. Koornneef (Wageningen) has identified a class of mutants (*hy*) that have long hypocotyls in the light. There are at least 7 loci which give the *hy* phenotype; 4 of these genes affect the activity of one or all of the red-light photoreceptors, the phytochromes. We have constructed double mutants between the phytochrome-deficient *hy* mutants and *det*1 and *det*2. The results from epistasis studies using doubly mutant lines suggest a hierarchical regulatory network among these genes in the control of the downstream light-regulated responses. Our results are consistent with a model where formation of the active form of phytochrome results in a decrease in activity of DET1 or DET2, which in turn leads to the induction of the light response. We are currently cloning *DET*1 and *DET*2 by chromosome walking using YAC libraries. Clones of the *DET* genes will lead to information on the mechanisms of the switch between light and dark growth modes in plants.

THE ROLE OF HOMEOBOX GENES IN MAIZE DEVELOPMENT
Sarah Hake[*], Ben Greene, David Jackson, Randy Kerstetter, Neelima Sinha, Laurie Smith, Bruce Veit. USDA/Plant Gene Expression Center, 800 Buchanan Street, Albany, CA 94710 and Department of Plant Biology, University of California, Berkeley, CA 94720 USA

The isolation and study of homeobox genes in animal systems has led to a detailed understanding of molecular events that regulate animal development. Members of homeobox gene families often act in concert to specify particular developmental processes. We discovered a family of homeobox genes in maize by transposon tagging one of the dominant *Kn1* mutations. *Kn1* mutations result in abnormal leaf morphology such that cells in the blade portion adopt fates of cells in the ligule or sheath regions. Clonal analysis suggested that the inner layer of the leaf was responsible for the mutant phenotype. More recent *in situ* localization shows that ectopic expression of the *Kn1* protein in the inner layer of the leaf is the probable cause of the mutant phenotype. The *Kn1* gene is normally expressed in the meristem and a subset of cells in shoot apex, but it is down regulated in leaves. Approximately ten homeobox genes have been isolated that are related to *Kn1* in the homeobox, but differ in sequence outside. We mapped the genes to chromosome position and analyzed expression patterns for some. Analysis of the *Kn1* mutations in maize and overexpression of the *Kn1* gene in tobacco allow us to make predictions of *Kn1*'s function, and to speculate on the role of plant homeobox genes in general.

SALICYLIC ACID - A NEW SIGNAL MOLECULE IN PLANTS

Ilya Raskin, Nasser Yalpani, Alexander J. Enyedi, Paul Silverman, & Yoram Kapulnik, AgBiotech Center, Cook College, Rutgers University, New Brunswick, NJ 08903-0231

Salicylic acid (SA) is a likely natural regulator of cyanide-resistant heat production in thermogenic inflorescences of *Arum* lilies and a systemic inducer of pathogenesis-related (PR) proteins associated with local and systemic disease resistance in plants. Both phenomena are associated with a dramatic increase of endogenous levels of SA, show structural specificity to this compound, and involve systemic translocation of SA. SA is formed from cinnamic acid and is rapidly converted to O-β-D-glucosyl SA in plants inoculated with necrotizing pathogens. Current results indicate that SA is an important signal molecule, which primarily acts at the level of gene expression.

PATTERN FORMATION DURING ARABIDOPSIS EMBRYO DEVELOPMENT.

U. Mayer, R. A. Torres Ruiz, T. Berleth, S. Miséra, G. Jürgens*

Pattern formation in the embryo generates the primary plant body organization as represented by the seedling. An apical-basal pattern consisting of epicotyl, cotyledons, hypocotyl and root is arranged along the single axis of polarity while a radial pattern perpendicular to the axis includes the main types of tissue such as epidermis, ground tissue and vascular tissue.

We have taken a genetic approach to analyze pattern formation in the Arabidopsis embryo. Following the isolation of a large number of mutants displaying abnormal seedling phenotypes, 9 genes have been identified that appear to be involved in three different aspects of body organization: apical-basal pattern, radial pattern, and shape. We are currently studying in detail how four of the genes contribute to the partitioning of the apical-basal axis in the early embryo.

BLUE LIGHT REGULATION OF GENE EXPRESSION
KA Marrs, KMF Warpeha, J Gao, K Bhattacharya, J Tilghman, J Marsh and LS Kaufman*. Department of Biological Sciences, University of Illinois at Chicago, Chicago, IL 60680, USA

The rate of transcription for the Cab gene family in Pisum sativum and Arabidopsis thaliana is affected by a single pulse of low fluence blue light. The effect is independent of protein synthesis, initiates soon after irradiation and only occurs for specific members of the Cab gene family. We are attempting to characterize the signal-transduction chain responsible for the blue-light induced transcription of these specific Cab genes. In the case of pea, one or several DNA binding proteins, present prior to blue-light irradiation, acquire the ability to recognize specific sequences in the regulatory regions of these specific Cab genes only after blue-light irradiation. Plasma membranes obtained from the apical buds of etiolated peas contain a flavoprotein responsible for activating a heterotrimeric GTP binding regulatory protein upon blue-light irradiation. The amount of light necessary to elicit flavoprotein excitation and subsequent G-protein activation is the same as that needed to elicit Cab gene transcription.

Hormonal Control of Plant Development. Harry Klee. Monsanto Company, 700 Chesterfield Village Parkway, St. Louis, MO 63198

Phytohormones have significant roles in controlling many aspects of plant development. Most of our knowledge of phytohormone action has been derived form exogenous application experiments. Such an approach has many disadvantages, including lack of control over uptake, transport and metabolism. In order to circumvent most of these difficulties, we have taken the approach of altering hormone levels in transgenic plants. We have isolated genes that can alter the levels of auxin, cytokinin or ethylene. Expression of these genes in plants has profound phenotypic effects on many aspects of plant development including apical dominance, vascular differentiation and fruit ripening. By combining different transgenes in a single plant, it has been possible to determine some of the biological roles for each of the hormones in isolation. The results of these experiments will be discussed.

SIGNAL PERCEPTION IN PLANTS: HEPTA-β-GLUCOSIDE ELICITOR BINDING PROTEINS IN SOYBEAN

Michael G. Hahn*, Jong-Joo Cheong, Robert M. Alba, and François Côté. Complex Carbohydrate Research Center & Departments of Botany and Biochemistry, University of Georgia, 220 Riverbend Road, Athens, GA 30602-4712, USA.

We are studying the cellular signaling pathway induced by a branched hepta-β-glucoside elicitor originally isolated from fungal walls. The hepta-β-glucoside elicitor is one example of an oligosaccharin (oligosaccharide with regulatory activities), a recently identified class of signal molecules active in plants. Treatment of soybean tissue with the hepta-β-glucoside elicitor leads to the induction of an important plant defense response, the biosynthesis and accumulation of antimicrobial phytoalexins. Structure-activity studies have identified key structural elements of the hepta-β-glucoside elicitor that are essential for its biological activity. Current research is focused on the first step in the signaling pathway, the recognition of the elicitor by a specific plant receptor. A radio-labeled derivative of the elicitor has been prepared and used to demonstrate the presence of specific, high-affinity binding protein(s) (EBPs) for the elicitor in microsomal membranes prepared from soybean roots. The EBPs co-migrate with plasma membrane markers in isopicnic sucrose density gradients. The EBPs have been solubilized from the soybean membranes using non-ionic detergents, and the solubilized proteins retain their high affinity and specificity for the hepta-β-glucoside elicitor. The EBPs recognize the same structural elements of the hepta-β-glucoside elicitor that are essential for its phytoalexin-inducing activity, suggesting that the EBPs are physiological elicitor receptors.

(Supported by NSF grant DCB-8904574. The Complex Carbohydrate Research Center is supported in part by the USDA/DOE/NSF Plant Science Centers Program through funding by DOE grant DE-FG09-87ER13810)

LEAFY CONTROLS MERISTEM IDENTITY IN *ARABIDOPSIS*

Detlef Weigel*, and Elliot M. Meyerowitz. Division of Biology, California Institute of Technology, Pasadena, CA 91125, USA

Development of a flower begins with the formation of a floral meristem by the inflorescence meristem. Mutations in the *Arabidopsis* gene *LEAFY* cause a partial transformation of flowers into inflorescence shoots, indicating that *LEAFY* promotes development of floral meristems as opposed to inflorescence meristems. Genetic studies show that *LEAFY* interacts with at least one other gene, *APETALA1*, in the determination of floral meristems. We have cloned the *LEAFY* gene, and, as a first step towards understanding the molecular basis of *LEAFY* function, we have analyzed its expression by *in situ* hybridization. Consistent with its role in determining floral meristem identity, *LEAFY* is expressed in floral, but not in inflorescence meristems. *LEAFY* expression precedes the expression of the floral homeotic genes *APETALA3* and *AGAMOUS*, whose expression patterns are altered in the absence of *LEAFY* activity. Furthermore, we found that *LEAFY* is the *Arabidopsis* homolog of the *Antirrhinum* gene *FLORICAULA*. *Antirrhinum* plants mutant for the *FLORICAULA* gene exhibit a more dramatic phenotype than *leafy* single mutants in *Arabidopsis*, i.e., a phenotype more similar to *leafy; apetala1* double mutants. The functional comparison of *LEAFY* and *FLORICAULA* reveals that species-specific differences exist in the basic mechanisms controlling flower development in dicotyledonous plants.

Integration of developmental signals during seed maturation in maize

Donald R. McCarty, Horticultural Sciences Dept. University of Florida, Gainesville, FL 32611.

During maturation of the seed embryo development is arrested and seed tissues destined to remain viable in the dry seed acquire dessication tolerance. Much of this developmental program is regulated through a seed specific response to the hormone abscisic acid (ABA). Most of the viviparous mutants of maize prevent maturation by blocking biosynthesis of ABA. The *vp1* mutant is exceptional in that it fails to respond to the hormone. In addition, *vp1* has a unique pleotropic effect of blocking activation of genes in the anthocyanin pathway. Genetic and molecular analyses indicate that *Vp1* regulates the anthocyanin pathway by activating transcription of a second regulatory gene, *C1* that in turn activates the downstream structural genes in the anthocyanin pathway. Comprehensive mutational analysis of the VP1 protein suggests that in addition to a general acidic transcriptional activation sequence VP1 contains functional domains required for activation of maturation genes that can be at least partially distinguished from domains required for activation of *C1* and the anthocyanin pathway. On the basis of these results we suggest that VP1 may function as a polyvalent activator capable of interacting with multiple transcription factors and that different interactions may be required for activation of different downstream genes. An important consequence of this activity may be the integration of an extrinsic hormonal signal with other intrinsic developmental signals specifying seed development. This is in accord with *vp1's* seed specific hormone insensitive phenotype.

MERISTEM INITIATION IN *ARABIDOPSIS THALIANA*
M. Kathryn Barton* and Scott Poethig. Department of Biology, University of Pennsylvania, Philadelphia, PA 19104, USA

We have described the development of the shoot and root apical meristems in both wild-type embryos and in embryos homozygous for mutations at the *shootless* and *rootless* loci. Seedlings homozygous for the *shootless* mutation lack a shoot apical meristem but are otherwise normal. Likewise, the main defect in seedlings homozygous for the *rootless* locus is the lack of a root apical meristem. Wild-type shoot apical meristem formation can be divided into two stages, a stratification stage and a broadening stage. The *shootless* mutation appears to block the transition between the stratification and broadening stages of meristem development.
The development of the wild-type root apical meristem is characterized by a stereotyped and partially invariant series of divisions. The *rootless* mutation causes an alteration in these division patterns such that the transition from simple patterns of division to more complex patterns is never realized. In addition, we describe the ability of *shootless* and *rootless* mutant tissue to produce shoots and roots postembryonically.

PERTURBING PETUNIA PIGMENTATION PATTERNS. Richard Jorgensen*. Departments of Vegetable Crops and Environmental Horticulture, University of California-Davis, Davis, Calif. 95616 USA

The familiar "picotee" and "star" patterns of garden petunias are due to control of chalcone synthase (CHS) expression in petal epidermal cells. White sectors in which CHS is not expressed are non-clonal and must be determined by intercellular communication. The mechanisms of CHS regulation and intercellular signalling are unknown. Analysis of pattern inheritance has indicated that pattern formation is controlled by multiple genes. A system in which such patterns can be elicited and perturbed genetically was discovered fortuitously. In petunia inbred lines which are stably and uniformly pigmented, introduction of a CHS transgene was found to produce a variety of non-clonal patterns which are due to co-suppression of the CHS transgene and the homologous, endogenous CHS gene in the white regions of the patterns. Interestingly, the transgene occasionally undergoes spontaneous epimutation to produce distinct new patterns. These new derivatives of the transgene are reversible and also may cause paramutation of the original transgene allele. Progeny testing of branches in which somatic epimutations of the transgene have occurred reveals that these events are germinally heritable. This is surprising in view of the fact that the pattern phenotype controlled by the transgene is expressed in L1-derived epidermal cells, while germ cells are L2-derived. This suggests that pattern elaboration in the epidermis is controlled by signals from mesophyll cells.

MOLECULAR GENETIC ANALYSIS OF FERTILIZATION IN *ARABIDOPSIS*.
R.E. Pruitt*, B.K. Kihl and S.E. Ploense. Department of Genetics and Cell Biology, University of Minnesota, St. Paul, MN 55108, USA

The male gametophyte of flowering plants has a very limited and highly specialized life-cycle. Following the maturation of the pollen grain, the growth and development of the male gametophyte is dictated largely by its cell-cell interactions with the female reproductive system. Recognition of the pollen grain by the stigmatic cells is required for germination of the pollen tube which is then dependent on other signals from the stigmatic cells for guidance and directional growth down into the transmitting tract. Once in the transmitting tract the pollen tube must find its way to an individual ovule and this process also appears to depend on signals from the female reproductive tissues. We have begun a genetic analysis of these various interactions and have succeeded in isolating a number of mutants which appear to disrupt some of these processes. These mutants are presently undergoing further characterization in the hope of elucidating some of the molecular processes involved in these regulatory cell-cell interactions.

AUXIN-BINDING PROTEINS AND THEIR POSSIBLE ROLE IN CELL DEVELOPMENT Alan M. Jones, Department of Biology, University of North Carolina, Chapel Hill, NC 27599, USA

The growth hormone class designated auxin is one of many signals in plants postulated to communicate position and change to cells. Induction of cell enlargement and differentiation along specific positions in the plant as well as wound-induced disruptions of auxin gradients leading to redifferentiation are notable examples of auxin effects. Among these effects, the molecular changes in cell enlargement have been the best characterized. Auxins induce changes in cell wall properties and specific gene transcription so rapidly that kinetic studies have been unable to discern which of these (or both) is the primary effect of this hormone and thus, it has been difficult to determine if these two essential processes are mediated by single or multiple receptor systems. With the recent discovery of several candidate auxin receptors, we must now seriously consider the possibility of multiple receptors. I will discuss the identification and characterization of two auxin-binding proteins (ABP) and the evidence for their role in development. One of these ABP's has unusual cellular trafficking to the plasma membrane/cell wall space. Another is located in the nucleus. I will explore the idea that the rapid changes that auxin causes to cell walls and to gene transcription are mediated separately by these two receptors.

ABSTRACTS
OF
SHORT TALKS

MUTATIONS AFFECTING PEA LEAF DEVELOPMENT
Strommer, JN, J Gerrath, and J Fausto
University of Guelph, Guelph, ON N1G 2W1, Canada

Pea leaves are complex organs, composed of a pair of basal stipules, one to three pairs of lateral leaflets and a terminal complex of tendrils. Tendrils can be converted to terminal leaflets, and leaflets to tendril complexes, each event the result of a single recessive mutation (tl and af, respectively; see Marx 1987). The mutant in which tendril complexes replace leaflets, af af, is of commercial interest, as the yield from semi-leafless plants approaches that of conventional peas (Snoad 1974) and the plants are more resistant to lodging and disease. As predicted from the success of af af plants in the field, tendrils are photosynthetically efficient exporters of fixed carbon (Côté et al. 1992a); as a consequence of the limited intercellular spaces in tendrils, these plants may also be more resistant to air pollution (Fausto and Strommer, unpublished).

Beyond the agronomic importance of af af plants, the Af and Tl genes are of interest as genes which function in development in such a way that inactivation of one causes substitution of either leaflet or tendril complex with the alternative structure. We are beginning a series of studies to learn how the Af gene acts in cooperation with Tl to program the development of normal leaves. There are three aspects of this work: an anatomical analysis to complement completed morphological studies of tendril development (Côté et al. 1992b), analysis of mutant sectors produced by loss of Af as a consequence of chromosome breakage, and cloning of the Af gene.

Molecular characterization of the AXR1 gene of *Arabidopsis thaliana*.
Ottoline Leyser, Cindy Lincoln, Candace Timpte, Jocelyn Turner, Doug Lammer and Mark Estelle. Indiana University, Department of Biology, Jordan Hall 142, Bloomington, IN 47405

We are taking a genetic approach to study the mode of action of the plant hormone auxin. We have used a variety of screens to identify genes involved in auxin action. One such gene, AXR1, is defined by a series of mutant alleles conferring resistance to normally inhibitory levels of exogenously added auxins. The mutations are all recessive and pleiotropic. Homozygous axr1 plants are dwarfed and show reduced apical dominance and defective root gravitropism. The degree of auxin resistance conferred by the different axr1 alleles correlates with the severity of other aspects of the phenotype. This implies that the AXR1 gene product is involved in auxin perception or response.

In the expectation of learning more about auxin action in plants, we have cloned the AXR1 gene by chromosome walking. We have identified a 16.5 kb DNA segment which, when transformed into axr1-3 mutant plants, rescues the mutant phenotype and restores wild-type levels of auxin sensitivity. By screening cDNA libraries and carrying out RNA blot analysis we have identified two transcripts from this region. The RNA blot analysis shows that one of these transcripts is not present in RNA extracted from plants homozygous for the radiation induced axr1-23 allele. We believe that this transcript represents the AXR1 mRNA and so we are currently sequencing both its cDNA and the genomic DNA from which it is transcribed.

Structure and Expression of Style-Expressed and Pollen-Expressed Components of Gametophytic Self-Incompatibility in *Petunia hybrida*. T.L. Sims, J.J. Okuley, K.R. Clark and P.D. Collins, Ohio State University, Columbus, OH 43210

Gametophytic self-incompatibility, a genetic barrier to inbreeding, is characterized by a localized interaction between pollen tubes and a stylar ribonuclease encoded by the S-locus. We have characterized the structure and expression of S-RNases in *Petunia hybrida*, and have shown that both temporal and spatial expression patterns are consistent with a role for these proteins in self-incompatibility. We have also characterized the expression patterns of several genes unrelated to the S-locus, that are preferentially expressed in pistils. Together with the S-RNases, these genes define several different expression programs that occur during development of the pistil. We are using transient assays, along with analysis of transgenic plants, to characterize those sequences required for proper developmental expression of the S-locus. Transient assays show that full-length promoter constructs are expressed in styles, ovaries, and petals, but not in leaves or pollen. Fluorometric assays of deletion constructs have provided preliminary evidence for a region conferring quantitative modulation of S-locus expression.

In contrast to the style-expressed S-RNase, the pollen component of gametophytic self-incompatibility remains uncharacterized. We have identified three sequences closely linked to the S_1-RNase gene that are expressed in other organs of *Petunia*. One of these sequences appears to encode a 2.2 kb transcript that accumulates to low levels in germinating pollen. Furthermore, the region encoding this transcript shows homology to sequence regions flanking the S_2 and S_3 alleles. We are using a variety of approaches to further characterize the S_1 pollen-expressed sequence, along with the homologous S_2 and S_3 sequence regions.

MOLECULAR CLONING AND CHARACTERIZATION OF A cDNA ENCODING A PEA HOMOLOG OF THE PROTEIN FARNESYLTRANSFERASE β SUBUNIT
Zhenbiao Yang[*], Carole L. Cramer[†], and John C. Watson[*‡]
[*]Department of Botany and [‡]Center for Agricultural Biotechnology, University of Maryland, College Park, MD 20742-5815
[†]Department of Plant Pathology, Physiology, and Weed Science, Virginia Polytechnic Institute and State University, Blacksburg, VA 24061

Protein farnesyltransferase is a heterodimeric enzyme that post-translationally modifies proteins such as p21[ras] by attaching a farnesyl moiety to a C-terminal cysteine residue. Genes encoding both the α and β subunits have recently been cloned and characterized from yeast and mammalian cells. To isolate a plant homolog, degenerate oligonucleotides, corresponding to the conserved regions of β subunit, were used as primers for polymerase chain reactions to amplify pea cDNA. Amplification of cDNA synthesized from total celluar RNA isolated from pea buds generated a 171 bp fragment with an open reading frame of 57 amino acids that shows 65% identity to the rat β subunit. Using the 171 bp fragment as hybridization probe to screen a pea cDNA library, one full-length cDNA clone was obtained that contains an open reading frame encoding a polypeptide of 419 amino acid. The predicted amino acid sequence exhibits 48% and 40% identity to the rat and yeast β subunits, respectively. RNA gel blot analysis shows that the pea homolog is expressed as a single transcript of approximately 1.7 kb in the buds of pea seedlings grown in complete darkness or in continuous white light.

ANALYSIS OF A RECEPTOR-LIKE PROTEIN KINASE OF *ARABIDOPSIS THALIANA*.
G. Eric Schaller, Sara Patterson, and Tony Bleecker
University of Wisconsin, Department of Botany, Madison, Wisconsin 53706

Receptor protein kinases play a central role in the cell-to-cell signalling of animal systems. Spanning the plasma membrane, they have an extracellular ligand-binding domain, a single transmembrane domain, and a cytoplasmic protein-kinase domain. Upon extracellular binding of ligand to receptor, the kinase domain in activated, generating an intracellular signal. Recently, genes have been cloned from higher plants which code for proteins structurally similar to the animal receptor kinases, suggesting that plants might have an analogous mechanism for signal transduction.

Genomic and cDNA clones encoding a novel receptor-like protein kinase (TMK1) have been isolated from *Arabidopsis thaliana*. The predicted protein has an intracellular kinase domain most related to the receptor tyrosine kinases but containing diagnostic serine/threonine sequences, a transmembrane domain, and an extracellular domain containing 11 copies of a leucine rich repeat. Leucine rich repeats have only been found in proteins associated with the plasma membrane and are involved in protein/protein interactions. Domain-specific antibodies against the extracellular and intracellular domains of TMK1 have been made using fusion proteins expressed in *E. coli*. In extracts of *Arabidopsis*, the antibodies specifically immunodecorate a polypeptide of about 120 kD. Digestion of the native protein from *Arabidopsis* with endoglycosidase F and neuraminidase reduces the apparent molecular mass of the immunodecorated protein by 10 kD, indicating that the native TMK1 protein is glycosylated. The intracellular domain of TMK1 was expressed as a fusion protein with maltose binding protein in *E. coli* and was found capable of autophosphorylation, indicating that the intracellular domain of TMK1 is a functional protein kinase.

LOVASTATIN INDUCES CYTOKININ DEPENDENCE IN TOBACCO CULTURES
Dring N. Crowell*, and Michael S. Salaz. Department of Biology, Indiana University-Purdue University at Indianapolis, Indianapolis, IN 46202-5132

De novo synthesis of mevalonic acid, which is catalyzed by HMG-CoA reductase, is the first committed step in the formation of isoprenoid compounds. Various studies have shown that mevalonic acid-derived compounds are required for cell growth (reviewed in 1). This conclusion is supported by the observation that cells treated with lovastatin (mevinolin), a potent inhibitor of HMG-CoA reductase, cease growth (1). We show that *Nicotiana tabacum* cells, which require exogenous auxin for growth in culture but do not require exogenous cytokinin, are growth-inhibited by 1 µM lovastatin. However, these cells are capable of growing in the presence of 1 µM lovastatin if 8 µM zeatin is supplied in the medium. Furthermore, benzyladenine, kinetin and thidiazuron are effective at reversing the inhibition of growth of these cells by 1 µM lovastatin, whereas adenine and 6-methyladenine have no effect. These results demonstrate that restoration of growth to lovastatin-treated tobacco cells is cytokinin-specific and is not caused by metabolism of cytokinin into other isoprenoid compounds. With the exception of kinetin, cytokinin does not effectively reverse the effects of higher concentrations of lovastatin (≥ 5 µM) but mevalonic acid does, consistent with the hypothesis that cytokinin biosynthesis is more sensitive to lovastatin than the biosynthesis of other essential isoprenoid compounds in tobacco cells. This observation suggests that lovastatin can be used to induce cytokinin-dependence in cytokinin-autonomous tobacco cell cultures and is, thus, a valuable tool for manipulating cytokinin levels *in vivo*.

[1] Goldstein, J. L., and Brown, M. S. (1990) Nature 343: 425-430.

GENE EXPRESSION DURING CYCLES OF GROWTH AND DORMANCY IN AXILLARY BUDS OF PEA
Joel Stafstrom, Plant Molecular Biology Center, Department of Biological Sciences, Northern Illinois University, DeKalb IL 60115

Axillary buds on intact pea seedlings (*Pisum sativum* L. cv. Alaska) do not grow. However, these buds quickly resume growing after decapitatng the terminal bud. Under appropriate conditions, growing buds can be made to stop growing and become dormant; these buds subsequently can grow again. Therefore, buds have the capacity to undergo multiple cycles of growth and dormancy. We have studied these growth cycles using expression of cDNA clone pGB8, which is homologous to ribosomal protein gene L27 from rat, as a molecular marker. pGB8 mRNA increases 13-fold 3 hr after decapitation, reaches a maximum enhancement of about 35-fold after 12 hr and persists at slightly reduced levels at later times. Terminal buds, root apices and elongating internodes also contain pGB8 mRNA but fully expanded leaflets and mature internodes do not. *In situ* hybridizations indicate that pGB8 expression increases in all areas of the bud within 1 hr of decapitation. Northern blots indicate that GB8 expression is reduced to dormancy levels as soon as buds stop growing. However, *in situ* hybridizations show that expression continues at growing-bud levels in the apical meristem for 2 days after it is reduced in the rest of the bud.

CALMODULIN ISOFORMS AND A CALMODULIN-LIKE Ca^{2+}-BINDING PROTEIN IN ARABIDOPSIS Ray Zielinski, Margaret Gawienowski, Dan Szymanski, & Imara Perera, Department of Plant Biology, University of Illinois, Urbana, IL 61801

One of the poorly understood aspects of Ca^{2+}-mediated signal transduction in plants is how the common signal of increased $[Ca^{2+}]_c$ is transduced to give specific physiological responses to different external stimuli. To investigate the possibility that differential stimulation of Ca^{2+}-mediated response pathways is directed by different EF-hand Ca^{2+}-receptor proteins in plants, we have isolated cDNA clones encoding representatives of calmodulin mRNAs from six different *Arabidopsis* genes, and a related sequence encoding a 22-kDa Ca^{2+}-binding protein (CaBP-22) that shares 65% amino acid sequence identity with calmodulin. The calmodulin cDNA sequences encode four different isoforms of the protein. ACaM-5 encodes a polypeptide identical to those encoded by ACaM-2 and 3, and corresponds to the thigmostimulated TCH-1 mRNA sequence (Braam and Davis, 1990. *Cell* **60**, 357). Expression levels of the Ca^{2+}-binding protein mRNAs in different tissues have been measured by slot blot hybridization and PCR amplification using gene-specific primers. ACaM-1, 4, and 5 mRNAs accumulated in all tissues tested, whereas ACaM-6 mRNA accumulated only in leaves. Similarly, CaBP-22 mRNA was detected primarily in leaves, but also accumulated at 10- to 100-fold lower levels in floral stalks and flowers. ACaM-2 and 3 mRNAs could not be detected in roots. These results suggest that the proteins encoded by these cDNA sequences play non-overlapping roles in regulating the physiological activities of different cell types. Current work is aimed at elucidating potential biochemical differences among the ACaM isoforms by expressing them in *E. coli*, from which they can be purified to homogeneity. We have also isolated the structural gene encoding the ACaM-3 calmodulin isoform. The promoter region of this gene, unlike animal calmodulin genes, is AT-rich and contains sequences resembling consensus heat shock and cyclic AMP response elements.

ABSTRACTS OF POSTERS

ISOLATION AND CHARACTERIZATION OF THE GENE ENCODING NODULE-ENHANCED GLUTAMATE SYNTHASE IN ALFALFA

Robert Gregerson[1,2], Susan Miller[2], Scott Twary[2], Steven Gantt[3], and Carroll Vance[1,2]. [1]Plant Science Research Unit, USDA-ARS, [2]Department of Agronomy and Plant Genetics and [3]Plant Biology Department, University of Minnesota, St. Paul, MN 55108.

Glutamate synthase (NADH-GOGAT) is one of several proteins whose expression is either enhanced in, or specific to developing legume root nodules induced by *(Brady)Rhizobium* infection. This enzyme catalyzes the second committed step in the assimilation of symbiotically fixed nitrogen; the transamidation of 2-oxoglutarate by glutamine to yield two moles of glutamate. We have previously purifed alfalfa root nodule NADH-GOGAT and raised antibodies against this protein. To investigate the regulation of the gene encoding this enzyme, we have isolated a full length cDNA clone for NADH-GOGAT from an alfalfa nodule cDNA library. This 7.5 kb cDNA includes the sequence encoding the mature 210 kd protein (the amino terminus of which has been sequenced), as well as an apparent transit peptide. Northern blots indicate that mRNA levels are developmentally regulated in the nodule, showing a similar pattern to that of previously characterized genes encoding aspartate aminotransferase and phosphoenoylpyruvate carboxylase. Protein levels as determined by Westen blots, and enzyme activity correlate well with mRNA levels. Southern blots suggest that NADH-GOGAT is encoded by a low or single copy gene. The deduced peptide encoded by this cDNA shows strong homology to both subunits of the *E. coli* glutamate synthase protein, and also to maize ferredoxin GOGAT.

Development of a Transient Expression System in *Arabidopsis* for Rapid Molecular Complementation of the Ethylene Response Locus (ETR).
Sara E. Patterson, Qianhong G. Chen, and Tony Bleecker
Botanty Department, University of Wisconsin, Madison, WI 53706

Techniques such as chromosome walking or gene tagging necessitate the use of molecular complementation to verify the identity of the cloned gene. Stable plant transformation techniques are often tedious and lengthy. As an alternative, rapid transient expression systems for molecular complementation have two requirements: 1) a method for introducing foreign DNA to be transiently expressed, and 2) a method for detection of complementation at the cellular level. We have developed a transient system designed to test overlapping genomic cosmid clones from a chromosome walk spanning the ETR-1 (ethylene response locus). For cellular level detection of the ethylene response we utilized transgenic plants containing a chitinase-GUS reporter fusion system in which GUS expression is ethylene inducible (Q. G. Chen). In etr1 background, plants are ethylene insensitive and GUS expression is inhibited.

For transient expression of cosmid clones we have investigated two methods : particle bombardment and transient expression via *Agrobacterium* infection. Particle bombardment experiments were carried out using the electrical discharge gun at Agracetus, Middleton, WI. Transient expression via *Agrobacterium* infection was studied using a 35S-Gus reporter gene containing a plant intron in the Gus coding sequence (35S-Gusint), which eliminates Gus expression by the bacterium. (Vancanneyt et al.). A variety of tissue types and *Arabidopsis* ecotypes were tested. Results of these experiments indicated that *Agrobacterium* infection of young stem segments resulted in reliable and high level transient expression of the 35S-Gusint reporter gene. The components of this system, taken together, should allow for efficient screening of cosmic clones from the chromosome walk and identification via molecular complementation of the etr1 mutation.

Q. G. Chen. see preceding abstract.
Vancanneyt, G, Schmidt, R., O'Connor-Sanchez, A., Willmitzer, L., and Rocha-Sosa, M. 1990 Mol. Gen. Genet. 220:245-250

GENETIC AND MOLECULAR STUDIES OF ETHYLENE RESPONSES IN *ARABIDOPSIS THALIANA*. Qianhong G. Chen and Tony Bleecker. University of Wisconsin, Department of Botany, Madison, Wisconsin 53706

We are investigating the molecular basis of ethylene responses in *Arabidopsis thaliana* by studying a number of ethylene-insensitive mutants. Some specific characteristics of individual alleles can be studied by producing ethylene dose-response curves. In addition to previously described dominant etr1-1 mutant(Bleecker, *et al.*,1988), we have measured the ethylene dose-response of hypocotyl growth in each mutant line. The mutation can be classified clearly into dominant, semi-dominant, or recessive based on this phenotype. The influence of gene dosage of specific alleles over wild-type was examined in F1 progeny from a cross of a homozygote diploid mutant line to a tetraploid wild-type line. In order to study the influence of etr mutations on ethylene-regulated gene expression, we have transformed *Arabidopsis* with an ethylene-responsive reporter gene. The GUS fusions with chitinase promoter pCG2226 from bean(Broglie, K.E., *et al.*, 1989) have been transferred into wild-type *Arabidopsis*. In all transgenic lines, GUS expression is induced by ethylene in leaf and stem tissue. In root tissue, GUS activity is constitutively expressed as a low level and is increased after exposure to ethylene. The reporter gene has been crossed into the mutant backgrounds to study the effects of different etr alleles on ethylene regulated gene expression. The expression of GUS is inhibited in all tissues of etr1-1 mutant plants.

1. Bleecker, A.B., Estelle, M.A., Somerville, C., and Kende, H. (1988) Science 241, 1086-1089.
2. Broglie, K.E., Biddle, P., Cressman, R., and Broglie, R. (1989) The Plant Cell 1, 599-607.

FACTORS THAT DETERMINE LONGEVITY AND SENESCENCE IN Arabidopsis thaliana
Linda Hensel, Michelle Nelson, David Baumgarten, Vojislava Grbic, Todd Richmond, and Tony Bleecker. Botany Department, University of Wisconsin, Madison

The maximum life span of the Arabidopsis plant is governed by two processes: a) the cessation of proliferative activity of the meristems and b) the loss of somatic tissues associated with senescence of leaves, stems, and siliques. The correlative control hypothesis for plant development postulates that senescence of somatic tissues is controlled by developmental signals associated with reproduction. We tested this hypothesis by measuring the timing of rosette leaf senescence in plants carrying single gene mutations which affect reproductive development. Results indicate that the timing of leaf senescence was not affected by mutations which delay flowering time or block fruit development. Instead, leaf senescence appeared to be an intrinsic, age-related process preceded by declining photosynthetic activity and declining expression of photosynthesis-related nuclear genes. On the other hand, proliferative activity of the inflorescence meristems appeared to be regulated by fruit development. The primary inflorescence meristem continued developing for twice as long and produced twice as many floral buds in the male-sterile mutant plants as compared to wild type control plants. New mutations which affect both the timing of leaf senescence and the cessation of meristematic activity are being isolated and will be described.

Genetic Characterization of Six *Arabidopsis thaliana* Late-Flowering Mutants.
Karin N. Lohman and Richard M. Amasino. Dept. of Biochemistry, UW-Madison.

Six *Arabidopsis thaliana* late-flowering mutants were characterized The late-flowering phenotype results from the extension of the vegetative phase of development and not a reduction in growth rate. Therefore, the late-flowering mutants form more leaves than wt. These mutations are monogenic and recessive and define four complementation groups that are allelic with previously identified late-flowering alleles. The phenotypic expression of two alleles may be modified by a dominant gene present in Arabidopsis ecotype Wassilewskija but not Landsberg erecta. The effect of vernalization and photoperiod on flowering time in these mutants was determined. Mutants 2307, 2804, and 3784 (allelic to gi) are slightly more sensitive to vernalization than wt and insensitive to photoperiod. Mutants 4136, 5459, and 7486, which are allelic to ld, fca, and fpa exhibit a strong response to vernalization and, when vernalized, also to photoperiod. These results are discussed in the context of multifactorial regulation of floral induction in *Arabidopsis thaliana*.

Genetic analysis of a heterochronic ecotype of Arabidopsis thaliana that has delayed conversion of lateral meristems from the vegetative to the reproductive phase

Vojislava Grbic and Anthony Bleecker, University of Wisconsin, Department of Botany, Madison

The shoot apical meristem of Arabidopsis thaliana goes through 3 phases during development: the phase 1 meristem is defined as a vegetative meristem that forms leaves organized into rosette; the phase 2 meristem is initial reproductive meristem that give rise to cauline leaves and lateral inflorescence meristem; the phase 3 meristem produces flowers. At the transition from vegetative to reproductive development in wild type Landsberg erecta (LER), the primary meristem and all lateral meristems are coordinately converted into phase 2 meristem. We report here on naturally occuring late flowering ecotype of Arabidopsis (SY-0) in which there is delayed conversion of the lateral meristems from vegetative to reproductive phase, resulting in aerial rosette formation at the nodes of the stem. Data from F1 crosses between SY-0 and LER, and analysis of segregating F2 and test cross populations reveal that the observed heterochronic phenotype is a result of two dominant genes that may be involved in signal processing during meristem development.

INDUCTION OF THE HYPERSENSITIVE RESPONSE IN TOBACCO BY *PSEUDOMONAS SYRINGAE* VIA A SIGNAL TRANSDUCTION PATHWAY
Merelee Atkinson, James Bina and Luis Sequeira. University of Wisconsin, Department of Plant Pathology, Madison, WI 53706.

Disease resistance in plants is typically associated with a hypersensitive response (HR) of plant cells that come in contact with the pathogen. This response leads to rapid plant cell death and is accompanied by the transcriptional activation of "defense" genes in neighboring cells. Together these events prevent spread of the infection. Induction of the HR in tobacco by *P. syringae* pathovars causes the sustained activation of a K^+ efflux/net H^+ influx response. The dependence of this response on calcium influx initially suggested the involvement of a signal transduction pathway. We report here that the K^+/H^+ response and calcium influx are accompanied by increased breakdown of phosphatidylinositol and by the accumulation of inositol 1,4,5-trisphosphate (IP_3). We do not know whether IP_3 accumulation is required for calcium influx and the K^+/H^+ response. However, inhibition of both by bromophenylacyl bromide, a nonspecific phospholipase inhibitor and by neomycin, an inhibitor of polyphosphoinositide metabolism, is consistent with this possibility. Staurosporine, a protein kinase inhibitor, and eicosatriynoic acid, a lipoxygenase inhibitor, block the K^+/H^+ response but have little effect on calcium influx, which suggests that there are signaling steps downstream of calcium influx. We propose that induction of the K^+/H^+ and hypersensitive responses proceeds through a complex signal transduction pathway.

A COMPLEX EXPRESSION PATTERN IS DIRECTED BY THE *ARABIDOPSIS* ACYL CARRIER PROTEIN A1 GENE PROMOTER IN TRANSGENIC TOBACCO
Scott R. Baerson and Gayle K. Lamppa The University of Chicago, Department of Molecular Genetics and Cell Biology, Chicago, IL 60637
The acyl carrier protein (ACP) plays a pivotal role in the *de novo* synthesis of fatty acids, which occurs primarily within plastids. During fatty acid biosynthesis, nascent chains are covalently bound to a phosphopantetheine prosthetic group attached at serine 38 of the protein. Two linked genes, A1 and A2, coding for nearly identical isoforms of the acyl carrier protein had previously been isolated from an *A.thaliana* (*columbia*) genomic library and sequenced. The amino acids deduced from the nucleotide sequence of the two genes indicate they encode distinct transit peptides, but the mature proteins are the same except for residue 79. A chimeric gene comprised of approximately 1.0 kb of the A1 gene 5' flanking sequence fused to the coding region of β-glucuronidase (GUS) was transformed into tobacco. Fluorimetric analysis showed that GUS activity levels were approximately 6-fold higher in roots than in young leaves, and GUS activity was greatly reduced in mature leaves. A comparison of GUS activity in etiolated and light-grown seedlings indicates that the A1 promoter is not tightly regulated by light. Histochemical analyses revealed a complex spatio-temporal expression pattern reflective of both the developmental status and growth rate of individual tissues. Deletion analyses are in progress to identify cis-acting elements involved in the transcriptional regulation of the ACP A1 gene.

An auxin resistant mutant of *Arabidopsis* is elongation deficient
Timpte, C., Wilson A., Green P.*, Liu Y.*, and Estelle, M. *Plant Gene Expression Center, MSU-DOE East Lansing MI 48824, Indiana University, Bloomington IN 47401

The *axr2* mutation in *Arabidopsis* confers resistance to auxin and dramatically affects development of the plant. Several aspects of the phenotype, including agravitropic inflorescences, shortened internodes, and the lack of etiolation, may be attributed to the failure of cells to elongate normally. We have taken two approaches to investigate this aspect of the phenotype of *axr2*. The first approach is to directly examine the ultrastructure of *axr2* by light microscopy and SEM. The dwarf phenotype results from both reduced cell elongation and reduced cell division in inflorescences and hypocotyls. The second approach examines transcription from the auxin responsive genes first identified in soybean, the SAUR genes (McClure and Guilfoyle 1978 Plant Mol. Biol. 9:611-623). The SAUR genes are primarily expressed in elongating tissue and SAUR transcripts accumulate rapidly in the presence of auxin (Gee, MA *et al.*, 1991 Plant Cell 3:419-430). A SAUR gene homolog has been cloned in *Arabidopsis* and designated SAUR-AC1 (Green, personal communication). The accumulation of SAUR-AC1 transcripts has been examined in wild-type and *axr2* plant tissues upon treatment with exogenous auxin. The expression of SAUR-AC1 in *axr2* plants is dramatically different from expression in wild-type plants in all tissues examined. These results suggest that the AXR2 gene product mediates a primary response to auxin through involvement in auxin-regulated genes.

CHARACTERIZATION OF A CHLOROPLAST HOLO-ACP SYNTHASE USING RECOMBINANT PROTEINS EXPRESSED IN *E. COLI*

Li-Ming Yang, Michael D. Fernandez, Gayle K. Lamppa, Dept. of Molecular Genetics and Cell Biology, University of Chicago, IL 60637

Holo-ACP synthase catalyzes the activation of acyl carrier protein (ACP), a key factor in *de novo* fatty acid synthesis. The activation reaction includes the transfer of a phosphopantetheine group from coenzyme A (CoA) to apo-ACP, producing holo-ACP (active form) and 3',5'-ADP. A chloroplast holo-ACP synthase was identified in this lab using an *in vitro* import assay and an organelle-free modification reaction. The enzyme was stimulated by CoA, inhibited by 3',5'-ADP and EDTA, and has an apparent molecular weight of 50 kD as determined by gel filtration chromatography. At present, we are in the process of purifying this enzyme from spinach chloroplasts. To facilitate the purification of the holo-ACP synthase, spinach ACP genes were overexpressed in *E. coli*. The recombinant form of the ACP precursor (preACP) is imported into chloroplasts *in vitro* and recognized by both a chloroplast processing enzyme and the chloroplast holo-ACP synthase. A gene coding for mature ACP, i.e. without the transit peptide, was also synthesized in *E. coli*. The radiolabeled recombinant ACP co-migrates with radiolabeled apo-ACP synthesized in a reticulocyte lysate system and is converted to holo-ACP by both an *E. coli* holo-ACP synthase and the chloroplast holo-ACP synthase.

TRANSIENT ß-GLUCURONIDASE EXPRESSION IN OAT PROTOPLASTS:
RNA and DNA ELECTROPORATION CONDITIONS AND RNA DEGRADATION
ANALYSIS
David C. Higgs and James T. Colbert, Department of Botany, Iowa State
University, Ames, IA 50011

Transient expression of the ß-glucuronidase (GUS) marker gene was used to optimize delivery of DNA and *in vitro* synthesized mRNA into oat (*Avena sativa L.*) suspension cell protoplasts. The goal was to establish a protoplast system to study plant RNA stability. Electroporation conditions (capacitance, voltage, and DNA amount) were investigated for their affect on transient GUS activity, and a time course of GUS activity post-electroporation was determined. An RNase protection assay was used to detect GUS RNA transcribed from electroporated DNA. After DNA electroporation, GUS RNA was detectable at 4 hr, was maximum at 12 hr, and had decreased by 24 hr. Electroporation of capped and non-capped *in vitro* synthesized GUS RNA showed that capping was required for translation in oat protoplasts. Degradation rates for capped and non-capped GUS RNAs were estimated by electroporating radiolabeled RNAs into oat protoplasts. Experiments using an RNA synthesis inhibitor to estimate GUS mRNA stability in oat protoplasts are in progress.

David L. Smith and Nina Fedoroff, Carnegie Institution of Washington, Department of Embryology, 115W University Pky., Baltimore, MD 21210

We are developing a transposon tagging system in *Arabidopsis thaliana* based on the *Activator* (*Ac*) element from maize. The system is based on two constructs. 1, An "anchored" *Ac* element driven by a CMV 35S promoter is used as a source of transposase. This construct is also contains a negative selection marker, *tms 2* from *Agrobacterium*. *Tms 2* codes for an indole acetamide hydrolase that confers sensitivity to plants grown on naphthalene acetamide. 2, An internally deleted *Ac* element referred to as *Ds*, is used as the mobilizable element. *Ds* contains a promoter/enhancer trap comprised of a promoterless or 35S core promoter containing-ß-glucuronidase (GUS) gene. *Ds* also includes *aph 4* which confers hygromycin resistance so that transpositions can be mapped in relation to the donor site. *Ds* is inserted between a mutated acetolactate synthase (ALS) coding sequence and a 35S promoter which after excision, provides resistance to a high level of the herbicide chlorsulfuron. Both constructs are marked by kanamycin resistance for transformation selection and were introduced into plants using an *Agrobacterium*-mediated root transformation and shoot regeneration protocol. Data will be presented from co-transformations where plants were regenerated from cells simutaneously transformed with both constructs and from cross pollination of plants containing separate constructs. Thus far, all of the selection markers work well and we will report on the excision and transposition efficiency of the system.

Identification and characterization of mutants of *Arabidopsis thaliana* with an altered response to auxin transport inhibitors. Max Ruegger, Lawrence Hobbie, Mark Estelle. Indiana University, Bloomington Indiana, 47405

Wild-type *Arabidopsis* roots grow at a reduced rate and fail to respond to gravity when exposed to polar auxin transport inhibitors. EMS- and DEB-mutagenized M_2 seeds were germinated on various auxin transport inhibitors and mutant seedlings which displayed altered root growth were recovered. We have tentatively grouped these mutants into three classes based upon their root phenotypes. Class I mutants have roots which are specifically resistant to chlorflurenol-methyl (CM) inhibition of elongation and gravitropism. In all other aspects, these mutants appear wild-type. Recent results suggest that the cause of this resistance may be the inability to hydrolyse the CM-ester to its free acid, which is thought to be the active form. Class II mutants have roots which are resistant to inhibition of elongation but not to inhibition of gravitropism. In contrast to the Class I mutants, these mutants appear to be resistant to a number of auxin transport inhibitors. A single mutant falls into Class III. Roots of this mutant display a hypersensitive response to auxin transport inhibitors. We expect that analysis of the Class II and Class III mutants will ultimately provide information about the molecular mechanism of polar auxin transport.

CALCIUM-DEPENDENT PROTEIN KINASE (CDPK) FROM *Arabidopsis thaliana*. Brad M. Binder* and Michael R. Sussman. Univ. of WI, Plant Sci. Building, 1575 Linden Drive, Madison. Wisconsin 53706, USA

Protein phosphorylation is known to control and modulate events in many cell types. We have examined a calcium-dependent protein kinase (CDPK) that is also regulated by lipids. Specifically, we have examined regions where this kinase autophosphorylates and investigated potential second messenger control of this kinase.

A full length cDNA encoding a CDPK has been isolated from *A. thaliana* and expressed as a fusion protein in *E. coli*. The CDPK portion of the fusion protein has an approximate molecular weight of 72 kDa as determined with SDS-PAGE and is 116 amino acid residues longer at the amino terminus than the prototype CDPKα gene previously identified in soybean.

The *E. coli*-expressed CDPK phosphorylates histone III-S in a calcium-dependent manner. Kinase activity was stimulated synergistically 16 - 30 fold by the addition of both calcium and 2 mg/ml crude lipid. Lipid stimulation was specific for lyso-phosphatidylcholine and phosphoinositides and did not occur with the addition of phosphatidylserine or phosphatidylcholine. It is not known if the observed lipid stimulation is a property of the native kinase with its *in vivo* substrates.

Like many protein kinases, the enzyme autophosphorylates and this activity is calcium regulated. To localize putative phosphorylation sites on the CDPK we constructed fusion proteins with different regions of the CDPK and tested them as kinase substrates. Our evidence suggests that the *E. coli*-expressed fusion protein containing the N-terminal domain of CDPK is a good substrate for the active kinase.

Characterization of a T-DNA Insertion Induced Auxin Resistant Mutant of *Arabidopsis*

Lawrence Hobbie and Mark Estelle, Dept. of Biology, Indiana University, Bloomington, IN 47405

An *Arabidopsis* mutant resistant to the root growth-inhibiting effects of auxin was recovered in a screen of the T-DNA transformant collection at DuPont, Inc. In preliminary experiments, the mutant plants do not show striking resistance to any other plant hormone tested. The aerial phenotype of the mutant appears wild-type. The only visible phenotype observed was a delayed gravitropic response in the roots; all previously isolated auxin-resistant mutants also show defects in gravitropism. However, genetic and phenotypic tests suggest that the mutant is different from previously characterized auxin-resistant mutants. Genetic analysis shows that the mutation cosegregates with kanamycin resistance, indicating that the mutation results from T-DNA insertion. In order to clone the affected gene, junction fragments from the insertion site have been isolated and are being used to identify the full-length gene in wild-type libraries. The characterization of this gene should lead to a better understanding of the molecular mechanism of auxin action.

NATURAL VARIATION IN 32 *ARABIDOPSIS THALIANA* WILDTYPES: RESPONSE TO PHOTOPERIOD AND VERNALIZATION

Björn Karlsson*, Gavin R. Sills and James Nienhuis. Department of Horticulture, Univ. of Wisconsin-Madison, Madison, WI 53706, USA

The use of *Arabidopsis thaliana* as a model plant for molecular genetics is widespread, however, utilization of the natural variation among ecotypes is limited. Among the primary environmental variables involved in the adaptation of ecotypes to different environments, photoperiod and vernalization and their interactions in induction to flowering are well documented. Nevertheless, the effects of these variables amongst a range of ecotypes does not exist.

By characterizing ecotypes for relative differences in response to photoperiod and vernalization, a broader range of materials maybe available for physiologic and genetic study. Within a 2 x 2 factorial set of treatments using 8 and 20 hour daylengths, as well as unvernalized and vernalized 24 days, 32 wildtypes were analyzed for the number of rosette, cauline and lateral release leaves at the time of flowering. The number of days to senescence of the first six basal rosette leaves was also recorded.

MULTIPLE SEQUENCES HOMOLOGOUS TO POLY(A) BINDING PROTEIN GENES IN ARABIDOPSIS THALIANA

P. Hilson*, J. Sedbrook, K. Carroll and P.H. Masson. Laboratory of Genetics, University of Wisconsin-Madison, Madison WI 53706, USA

How plant mRNAs get their polyadenylate tail and what roles it plays is poorly understood. Animal and yeast mRNAs in vivo are found as ribonucleoproteins in which the major component is the poly(A) binding protein (PABP). The functions of the poly(A) tail rely on the formation of the poly(A)-PABP complex. The PABP is thought to participate in the control of gene expression by regulating the stability and the translational competence of mRNAs.

To gain more insight into the structure and possible functions of the PABP in plants, we have PCR-amplified and cloned two *Arabidopsis thaliana* genomic fragments (PABA1 and PABA2) homologous to a RNA recognition consensus present in the yeast, human, *Xenopus* and *Drosophila* PAB genes. Southern blot analyses show that these two PCR products correspond to distinct unique sequences in the *Arabidopsis* genome and hybridize to additional bands in medium stringency conditions.

We have isolated three clones homologous to the PABA1 probe from an *Arabidopsis thaliana* genomic library. Preliminary sequence data and comparison to available PAB gene sequences indicate that two of them encode PABP polypeptides while the third is probably a pseudogene. Our results suggest that a PAB ancestor gene gave rise to a plant multigene family. This complexity could underlie uninvestigated regulation mechanisms if different plant PAB genes have specific functions or specific patterns of expression.

CHARACTERIZATION OF GROWTH-SPECIFIC cDNA CLONES FROM PEA AXILLARY BUDS

Michelle Devitt and Joel Stafstrom, Plant Molecular Biology Center, Department of Biological Sciences, Northern Illinois University, DeKalb IL 60115

Axillary buds of intact Pea seedlings are dormant, but after decapitation begin to grow rapidly. This study characterizes genes whose expression is increased during this developmental transition. Several clones were isolated from a growing bud cDNA library. These clones were sequenced and their patterns of expression analyzed by RNA gel blotting. Clones that could be identified by sequence similarity include homologs of ribosomal protein genes L27 (clone pGB8) and L34 (pGB6) and histone H4 (pGB21). Twenty-four hours after decapitation, the relative expression of these genes increased 23-fold, 6-fold and 10-fold, respectively. We also looked at the expression of clones obtained from other labs. The expression of two IAA-responsive genes, pIAA4/5 and pIAA6 (from A. Theologis) showed increases of 23-fold and 4-fold after 24 h. The expression of pEA207 (from M. Dobres) does not change significantly in axillary buds following decapitation. We are characterizing other growth-specific clones and obtaining additional clones corresponding to members of histone and ribosomal protein multigene families.

Hyperproliferation of adventitious and lateral roots in a putative auxin mutant of *Arabidopsis thaliana*

J.J. King and D.P. Stimart. Department of Horticulture. University of Wisconsin, 1575 Linden Dr., Madison, WI 53706

Auxin-induced responses in plants can be regulated by endogenous IAA concentrations or by differential tissue specific sensitivity to auxin. Initiation of adventitious roots is a uniquely auxin-dependent developmental process. Variation in competence of stem cuttings to form adventitious roots is quantified as the number of roots formed per cutting and the percentage of cuttings forming roots. To examine the genetic basis of this variation, an EMS-mutagenized population of *Arabidopsis thaliana* 'Columbia' was screened for mutants with altered abilities to form adventitious roots. The phenotype of a selected recessive mutation, *rooty (rty)*, is characterized by overproduction of adventitious and lateral roots and inhibition of normal shoot development. In addition, dark-grown seedlings partially exhibit the ethylene associated triple response: hypocotyl and root elongation is inhibited and radial expansion of the hypocotyl occurs, but no apical hook forms. A phenocopy of the *rty* mutant develops when wild type seedlings are grown on medium containing 100 µM IBA (indole-3-butyric acid), and partial normalization of mutant seedlings occurs on medium containing the cytokinin 2iP (N^6-[2-isopentyl]adenine) at 12 to 25 µM. Root cultures of wild type require exogenously applied auxin to sustain growth; however, *rty* root cultures maintain slow, highly branched growth in auxin free medium. From these observations it appears that elevated endogenous IAA concentrations or increased sensitivity to auxin results from homozygosity of the mutant *rty* allele.

CLUES TO STOMATAL PATTERNING IN *TRADESCANTIA*

Judith Croxdale[*], Jonathan Chin, John Boetsch, Brian Yandell, John Goodman. Departments of Botany and Statistics, University of Wisconsin, Madison, WI 53706

Arrested stomatal initials represent loss-of-function events and were used as clues to understand stomatal patterning in *Tradescantia* leaves. The spatial distribution and position of arrested cells relative to stomata were studied. The distribution of arrested cells on half a leaf blade is clustered, while stomata are regularly distributed. On a two-dimensional basis, arrested cells are closer to stomata than stomata are to one another. Within cell files (a linear or one dimensional basis), arrested cells occur midway between the nearest stomatal pair at leaf maturity, but have no consistent position in relation to stomata in the area of origin. The number of epidermal cells that separate arrested cells from their nearest stomatal pair is the same as the number of cells that separate equivalent stomata. Spacing of arrested cells and stomata is not modified following their origin. Based on their spatial distribution and position relative to stomata, the results indicate that cell-cell interactions may occur between cell files, but not within them, that lead to the arrest of stomatal initials.

GLOBULIN GENE EXPRESSION IN MAIZE EMBRYOS

Renato Paiva and Alan L. Kriz
University of Illinois, Department of Agronomy

The products of *Globulin* genes represent the major storage proteins found in maize embryos. The genes *Globulin-1* (*Glb-1*) and *Globulin-2* (*Glb-2*) which encode storage proteins of Mr 63,000 and Mr 45,000, respectively, accumulate to high levels during normal embryo development and are degraded during the early stages of seed germination. Embryos of maize *viviparous* (*vp*) mutants fail to complete maturation but rather undergo precocious germination while attached to the ear. Our lab has been interested in investigating the effects of this precocious germination on the expression of *Globulin* genes. Our previous analysis of *Glb-1* transcript levels in *vp* embryos indicates that these embryos do not fully switch from a program of maturation to one of germination. This work also revealed that *Glb-1* is positively regulated by ABA and that expression of this gene requires a functional *Vp1* gene product. With the exception of the *vp1* mutant, which is non-responsive to ABA, embryos of the *vp* mutants are defective in ABA synthesis and therefore contain subnormal levels of this growth regulator. We collected embryos of normal and *vp* kernels prior to and during the onset of precocious germination. For each of these samples, RNA was extracted for analysis of transcript levels by northern blots and ABA levels were determined by GC-MS assay. This study indicates that *Globulin* expression is under the control of different factors which at least require a combination of *in situ* ABA levels and the presence of a functional *Vp1* gene product. Results of this study will be presented and discussed.

PARTICIPANTS

Ahmed L. Abdel-Maywood
Department of Plant Pathology
University of Wisconsin-Madison
1630 Linden Dr., Russell Labs
Madison, WI 53706

Scott Atkins
Department of Plant Biology
University of Wisconsin-Madison
1630 Linden Drive, Russell Labs
Madison, WI 53706

Merelee Atkinson
Plant Pathology Department
University of Wisconsin-Madison
1630 Linden Drive, Russell Labs
Madison, WI 53706

Scott Baerson
Department of Molecular
  Genetics & Cell Biology
University of Chicago
920 E. 58th Street - Box 16
Chicago, IL 60637

Pauline Bariola
MSU-DOE Plant Research Laboratories
Michigan State University
322 Plant Biology
East Lansing MI 48824

Hilla Ben-David
Department of Plant Biology
University of California-Berkeley
211 GPBB
Berkeley, CA 94720

Greg Bertoni
Department of Botany
University of Wisconsin-Madison
B101 Birge Hall
Madison, WI 53706

Mary Bett
Department of Plant Biology
University of Wisconsin-Madison
1630 Linden Drive, Room 789A
Madison, WI 53706

Keshab Bhattacharya
Department of Biological Sciences
University of Illinois-Chicago
Laboratory for Cell, Molecular and
  Developmental Biology
Chicago, IL 60680

Dinakar Bhattramakki
Department of Agronomy
University of Illinois
1201 W. Gregory Dr., 240 PABL
Urbana, IL 61801

James Bina
Department of Plant Pathology
University of Wisconsin-Madison
1630 Linden Dr, 885 Russell Labs
Madison, WI 53706

Brad Binder
Department of Horticulture
University of Wisconsin-Madison
Room 309
Madison, WI 53706

Milton Bruening
Department of Biology
Northwest Missouri State
205 GS Building
Maryville, MO 64468

Barry Brule
Department of Botany
University of Wisconsin-Madison
B119 Birge Hall
Madison, WI 53706

Mary Beth Carter
Department of Biochemistry
University of Wisconsin-Madison
420 Henry Mall
Madison, WI 53706

Amarjit Chahal
Department of Horticulture
  Science
University of Guelph
Guelph, Ontario N1G 2W1
CANADA

Q. Grace Chen
Department of Botany
University of Wisconsin-Madison
430 Lincoln Drive, Birge Hall
Madison, WI 53706

Diane Church
Department of Botany
University of Wisconsin-Madison
430 Lincoln Drive, Birge Hall
Madison, WI 53706

Dr. Dring N. Crowell
Department of Biology
Indiana University-Purdue
723 W. Michigan Street
Indianapolis, IN 46202-5132

Judith Croxdale
Department of Botany
University of Wisconsin-Madison
430 Lincoln Drive, Birge Hall
Madison, WI 53706

Kent Cushman
Department of Horticulture
University of Wisconsin-Madison
1575 Linden Drive
Madison, WI 53706

Steve Daniel
Department of Botany
University of Wisconsin-Madison
430 Lincoln Drive
Madison, WI 53706

Jack Dekker
Department of Agronomy
Iowa State University
3214 Agronomy Hall
Ames, IA 50011

Michelle Devitt
Department of Biological Sciences
Northern Illinois University
DeKalb, IL 60115

Stan Duke
Department of Agronomy
University of Wisconsin-Madison
246 Moore Hall
Madison, WI 53706

Philip Dykema
Department of Botany
Iowa State University
443 Bessey Hall
Ames, IA 50011

Larry Ennis
University of Wisconsin-Madison
845 Hemlock Drive
Verona, WI 53593

Dr. Kenneth Eskins
Seed Biosynthesis Research Unit
USDA, ARS, NCAUR
1815 N. University Street
Peoria, IL 61604

Joy Fausto
Department of Molecular
  Biology and Genetics
University of Guelph
Guelph, Ontario N1G 2W1
CANADA

Donna Fernandez
Department of Botany
University of Wisconsin-Madison
Birge Hall
Madison, WI 53706

Keven Folta
Department of Biological Sciences
University of Illinois-Chicago
Chicago, Il 60680

Janice Fullenwider
Department of Biochemistry
University of Wisconsin-Madison
420 Henry Mall
Madison, WI 53706

Susheng Gan
Department of Biochemistry
University of Wisconsin-Madison
420 Henry Mall
Madison, WI 53706

Jie Gao
Department of Biological Sciences
University of Illinois-Chicago
Laboratory for Cell, Molecular, and
  Developmental Biology
Chicago, IL 60680

Margaret Gawienowski
Department of Plant Biology
University of Illinois
190 PABL
Champaign, IL 61820

Pedro Gil
Department of Plant Research
Michigan State University
322 Plant Biology
East Lansing, MI 48824

Robert M. Goodman
Department of Plant Pathology
University of Wisconsin-Madison
1630 Linden Drive, 687 Russell Labs
Madison, WI 53706

Sherrie Gore
Department of Biochemistry
University of Wisconsin-Madison
420 Henry Mall
Madison, WI 53706

Robert Gregerson
Department of Agronomy and
  Plant Genetics
University of Minnesota
1991 Buford Circle, Borlaug Hall
St. Paul, MN 55108

Vojislava Grbić
Department of Genetics
University of Wisconsin-Madison
430 Lincoln Drive, B132 Birge Hall
Madison, WI 53706

Milton Haar
Department of Agronomy
Iowa State University
Ames, IA 50011

William James Hahn
Department of Botany
University of Wisconsin-Madison
Birge Hall
Madison, WI 53706

Jim Harbage
Deparment of Horticulture
University of Wisconsin-Madison
1575 Linden Drive
Madison, WI 53706

Gregory R. Heck
Department of Botany
University of Wisconsin-Madison
430 Lincoln Drive, 132 Birge Hall
Madison, WI 53706

Linda Hensel
Department of Botany
University of Wisconsin
430 Lincoln Drive, B137 Birge Hall
Madison, WI 53706

Cynthia Henson
Department of Agronomy
University of Wisconsin-Madison
345 Horticulture-Moore Hall
Madison, WI 53706

David Higgs
Department of Botany
Iowa State University
353 Bessey Hall
Ames, IA 50011

Pierre Hilson
Laboratory of Genetics
University of Wisconsin-Madison
445 Henry Mall
Madison, WI 53706

Lawrence Hobbie
Department of Biology
Indiana University
Jordan Hall
Bloomington, IN 47405

Barb Holman
Forest Products Laboratory
USDA
One Gifford Pinchot Drive
Madison, WI 53705-2398

Charles Hoogstraten
Department of Biochemistry
Univerity of Wisconsin-Madison
420 Henry Mall
Madison, WI 53706

Rebecca Hoogstraten
Department of Biochemistry
University of Wisconsin-Madison
1630 Linden Drive, Russell Labs
Madison, WI 53706

Christie Howard
Department of Plant Research
Michigan State University
322 Plant Biology Building
East Lansing, MI 48824

Glen Howe
Department of Forest Resources
University of Minnesota
115 Green Hall
St. Paul, MN 55108

Estelle Hrabak
Department of Horticulture
University of Wisconsin-Madison
1575 Linden Drive
Madison, WI 53706

Melissa Hurley
Department of Botany
University of Wisconsin-Madison
430 Lincoln Drive, 113 Birge Hall
Madison, WI 53706

Emily Jordan
Department of Horticulture
University of Wisconsin-Madison
1575 Lincoln Drive, Room 495
Madison, WI 53706

Björn Karlsson
Department of Horticulture
University of Wisconsin-Madison
1575 Linden Drive
Madison, WI 53706

Kenneth Keegstra
Department of Botany
University of Wisconsin-Madison
B121 Birge Hall
Madison, WI 53706

Joseph King
Department of Horticulture
University of Wisconsin-Madison
1575 Linden Drive
Madison, WI 53706

Steve Kopczak
Department of Genetics and
  Cell Biology
University of Minnesota
1445 Gortner Ave., 250 Biosciences Gr.
St. Paul, MN 55108

Doug Lammer
Department of Biology
Indiana University
Jordan Hall
Bloomington, IN 47405

Ilha Lee
Department of Biochemistry
University of Wisconsin-Madison
420 Henry Mall, Room 270
Madison, WI 53706

Ottoline Leyser
Department of Biology
Indiana University
162 Jordan Hall
Bloomington, IN 47405

Baochun Li
Department of Horticultural
  Sciences
University of Guelph
Guelph, Ontario N1G 2W1
CANADA

Karin N. Lohman
Department of Horticulture
University of Wisconsin-Madison
1575 Linden Drive
Madison, WI 53706

Dingwei Lu
Department of Botany
University of Wisconsin-Madison
430 Lincoln Drive, 113 Birge Hall
Madison, WI 53706

Barbara MacGregor
Department of Bacteriology
University of Wisconsin-Madison
1550 Linden Drive
Madison, WI 53706

John Marsh
Department of Biological Sciences
University of Illinois-Chicago
Chicago, IL 60680

Amy Masshardt
Department of Biochemistry
University of Wisconsin-Madison
420 Henry Mall
Madison, WI 53706

Patrick Masson
Laboratory of Genetics
University of Wisconsin-Madison
445 Henry Mall
Madison, WI 53706

Dennis Mathews
Department of Biochemistry
University of Wisconsin-Madison
420 Henry Mall
Madison, WI 53706

Barry Micallef
Department of Botany
University of Wisconsin-Madison
430 Lincoln Drive, Birge Hall
Madison, WI 53706

Michelle Nelson
Department of Botany
University of Wisconsin-Madison
B139 Birge Hall
Madison, WI 53706

James Nienhuis
Department of Horticulture
University of Wisconsin-Madison
1575 Linden Drive
Madison, WI 53706

Ron Okagaki
University of Florida
Gainesville, FL 32606

Tom Osborn
Department of Agronomy
University of Wisconsin-Madison
Madison, WI 53706

Renato Paiva
Department of Agronomy
University of Illinois
240 PABL - 1201 West Gregory Dr.
Urbana, IL 61801

Sara E. Patterson
Department of Botany
University of Wisconsin-Madison
430 Lincoln Drive, B137 Birge Hall
Madison, WI 53711

Sharyn Perry
Department of Botany
University of Wisconsin-Madison
430 Lincoln Drive, 132 Birge Hall
Madison, WI 53706

Sara Ploense
Department of Genetics and
   Cell Biology
University of Minnesota
1445 Gotner Ave., 250 Biosciences
   Center
St. Paul, MN 55108

Joseph Pomerening
Department of Biochemistry
University of Wisconsin-Madison
420 Henry Mall
Madison, WI 53706

Patchara Pongam
Department of Plant Pathology
University of Wisconsin-Madison
1630 Linden Drive
Madison, WI 53706

Dr. H. Presley
2100 North Hudson
Chicago, IL 60614

Kathryn Richmond
Department of Plant Biology
University of Wisconsin-Madison
1630 Linden Drive, Russell Labs
Madison, WI 53706

Todd Richmond
Department of Genetics
University of Wisconsin-Madison
430 Lincoln Drive
Madison, WI 53706

Max Ruegger
Department of Biology
Indiana University
Jordan Hall
Bloomington, IN 47405

Holly J. Schaeffer
Department of Plant Pathology
c/o Dr. Martha Hawes
University of Arizona
Tucson, AZ 85726

Eric Schaller
Department of Botany
University of Wisconsin-Madison
430 Lincoln Drive
Madison, WI 53706

Katherine Schmid
Department of Biological Sciences
Butler University
4600 Sunset Ave.
Indianapolis, IN 46208

Rebecca Schneider-Brockman
Department of Biochemistry
University of Wisconsin-Madison
420 Henry Mall
Madison, WI 53706

Gavin Sills
Department of Horticulture
University of Wisconsin-Madison
1575 Linden Drive
Madison, WI 53706

Thomas L. Sims
Department of Biological Sciences
Ohio State University
406A Hottman Hall
Columbus, OH 43210

David Smith
Department of Embryology
Carnegie Institution of Washington
115 W. University Parkway
Baltimore, MD 21210

Joel Stafstrom
Department of Biological Sciences
Northern Illinois University
DeKalb, IL 60115

Susan Stieve
Department of Horticulture
University of Wisconsin-Madison
1575 Linden Drive
Madison, WI 53706

Julie Stone
Department of Horticulture
University of Wisconsin-Madison
1575 Linden Drive, Room 431
Madison, WI 53706

Judith Strommer
Department Horticultural Science
  and Molecular Biology and Genetics
University of Guelph
Guelph, Ontario N1G 2W1
CANADA

Michael Sullivan
Department of Horticulture
University of Wisconsin-Madison
1575 Linden Drive
Madison, WI 53706

Thomas Sullivan
Laboratory of Genetics
University of Wisconsin-Madison
445 Henry Mall
Madison, WI 53706

Dan Szymanski
Department of Plant Biology
University of Illinois
1201 W. Gregory, 356 PABL
Urbana, IL 61801

Judi Tilghman
Department of Biological Sciences
University of Illinois-Chicago
Chicago, IL 60680

Candice Timpte
Department of Biology
Indiana University
Jordan Hall
Bloomington, IN 47405

Baishnab C. Tripathy
Department of Botany
University of Wisconsin-Madison
430 Linden Drive
Madison, WI 53706

Ambro vanHoof
Plant Research Lab/Genetics
Michigan State University
322 Plant Biology
East Lansing, MI 48824

Richard Vierstra
Department of Horticulture
University of Wisconsin-Madison
1575 Linden Drive
Madison, WI 53706

Liwen Wang
Department of Biological Sciences
University of Illinois-Chicago
834 S. Miller 1st Floor
Chicago, IL 60607

Leigh Weber
Department of Horticulture
University of Wisconsin-Madison
1575 Linden Drive
Madison, WI 53706

Laurie Weiss
Department of Horticulture
University of Wisconsin-Madison
1575 Linden Drive, Room 431
Madison, WI 53706

David Wolyn
Department of Horticultural Science
University of Guelph
Guelph, Ontario N1G 2W1
CANADA

Liming Yang
Department Molecular Genetics and
  Cell Biology
University of Illinois-Chicago
920 E. 58th Street
Chicago, IL 60637

Zhenbiao Yang
Department of Botany
University of Maryland
3234 H. J. Patterson Hall
College Park, MD 20742

Ray Zielinski
Department of Plant Biology
University of Illinois
1201 W. Gregory Dr., 190 PABL
Urbana, IL 61801

AUTHOR INDEX

Alba, R. M.   7, 144
Altschmied, Lothar   57
Amasino, R. M.   158
Atkinson, M.,   160

Baerson, S.   160
Barton, M. K.   69, 145
Baumgarten, D.   156
Berleth, T.   93, 142
Bhattacharya, K.   21, 143
Bina, J.   160
Binder, B.   163
Bleecker, A. B.   123, 131, 152, 156, 159
Boetsch, J.,   166

Cabrera, H.   57
Carroll, K.   165
Chen, Q. G.   156
Cheong, J.-J.   7, 144
Chin, J.   166
Chory, J.   57, 141
Clark, K. R.   105, 151
Colbert, J. T.   162
Collins, P. D.   105, 151
Côté, F.,   7, 144
Cramer, C. L.   151
Crowell, D. N.   37, 152
Croxdale, J.   166

Devitt, M,   165

Enydei, A. J.   142
Estelle, M.,   150, 161, 163, 164

Faust, J.   150
Fedoroff, N.   162
Fernandez, M. D.,   161

Gantt, S.   156
Gao, J.   21, 143
Gawienowski, M.   153
Gerrath, J.   150
Goodman, J.   166
Grbic´, V.   156, 159
Greene, B.   141, 161
Gregerson, R.   156

Hahn, M. G.   7, 144
Hake, S.   63, 141
Hensel, L.   123, 156
Higgs, D.   162
Hilson, P.   165
Hobbie, L.   163, 164

Jones, A. M.   1, 147
Jorgensen, R.   87, 146
Jurgens, G.   93, 142

Kapulnik, Y.   142
Karlsson, B.   164
Kaufman, L. S.   21, 143
Kerstetter, R.   141
Kihl, B. K.   146
King, J. J.   166
Klee, H.   45, 143
Kriz, A.L.   167

Lammer, D.   150

Lamppa, G. K. 160, 161
Leyser, O. 150
Li, H.-m. 57
Lincoln, C. 150
Linzer, Rebecca A. 15
Liu, Y. 161
Lohman, K. N. 158

Marrs, K. A. 21, 143
Marsh, J.III 21, 143
Masson, P. H. 165
Mayer, U. 93, 142
McCarty, D. R. 27, 145
Meyerowitz, E. M. 115, 144
Miller, S. 156
Min-Wong, Lu 51
Miséra, S. 93, 142

Nelson, M. 158
Niehuis, J. 164

Oeller, Paul W. 51
Okuley, J. J. 105, 151

Paiva, R. 167
Patterson, S. E. 131, 152, 156
Perera, I. 153
Ploense, S. E. 146
Poethig, S. 145
Pruitt, R. E. 146

Raskins, I. 15, 142
Richmond, T. 158
Romano, Charles 45
Ruegger, M. 163
Ruiz, R. A. T. 93, 142

Salaz, M. S. 37, 152

Schaller, E. 131, 152
Sedbrook, J. 165
Sequeira, L. 160
Sills, G. R. 164
Silverman, P. 15, 142
Simon Misèra 93
Sims, T. L. 105, 151
Sinha, N. 141
Smith, D. 162
Smith, L. 141
Stafstrom, J. 75, 153, 165
Stimart, D.P. 166
Strommer, J. 150
Susek, Ronald 57
Sussman, M. R. 163
Szymanski, D. 153

Theologis, A. 51
Tilghman, J. 21, 143
Timpte, C. 150, 161
Turner, J. 150
Twary, S. 156

Vance, C. 156
Verbeke, J. A. 99
Viet, B. 141

Warpeha, K. M. F. 21, 143
Watson, J. C. 151
Weigel, D. 115, 144
Wilson, A. 161

Yalpani, N. 142
Yandell, B. 166
Yang, L.-M. 161
Yang, Z. 151

Zielinski, R. 153

SUBJECT INDEX

ABA 45
Abscisic acid 28, 127
ACC deaminase 46, 53
ACC oxidase 51
ACC synthase 51
Acidification 2
Adh 61
Agamous 118
Agrobacterium tumefaciens 43, 45
Allyl alcohol 61
Altered phyllotaxis 70
Alternative oxidase 16
Alternative pathway 16
An1 87
An2 87
Anthocyanin 29, 75
Antisense RNA 52
Apetala1 118
Apetala2 118
Apical cell 94
Apical dominance 47, 75
Apical meristem 97, 75
Arabidopsis 123, 131
Arum 15
Auxin 2, 45,
Auxin binding proteins 3
Axillary buds 75
axr-1 mutant 80
axr2 45
Azidoindole-3-acetic acid 3

Basal cell 94
Blue light responses 21
Brassica 136

C1 29
Cab 21
Cadastral genes 118

Calorigen 15
Carpels 100, 117
Catharanthus roseus 100
Cell cycle 78
Cell division 43
Chalcone synthase 88
Cholera toxin 23
Chlorophyll a/b binding protein 58
Chloroplasts 57
Clonal sectors 90
Co-supression 88
Corpus 69
Crown gall 45
Crucifers 94
Cytokinins 37, 45, 75, 125

Desiccation 27
Det 60
Differentiation 99
Disease resistance
Doe 60
Drought stress 28
Dwarf 64

ein1 47
ein2 47
Elicitors 7
Embryo 93
Embryogenesis 27
Epidermal 95
Epigenetic 90
Epimutation 76
Epinastic 48
Ethylene 45, 51, 80, 125

Fackel 95
Fass 97

Flavin 23
Floral induction 115
Floral primordia 120
Flower 87
Fruit ripening 48

G-protein 23
Gibberellic-acid 127
Gnom 95
Gurke 95

Häuptling 97
High-fluence 23
Hepta-β-glucosides 8
Heterochronic 117
Histone 78
HMG CoA reductase 37
Homeodomain 64
hsp70 46
Hygromycin 60
Hyperpolarization 2
Hypersensitive response 16
Hypophysis 71

IAA 45
iaaH 45
iaaL 46
iaaM 45
Indoleacetamide hydrolase 46
Inflorescence 115
ipt 46
ipt gene 80
Isopentenyl transferase 46
Isoprenoid 40

Keule 96
KN1 63
Knolle 96
Knotted 67, 99

Lateral buds 47
Laterne 97
Leaf 48
LEAFY 116
Lectin 79
Leucine rich repeats 133
Ligand binding 12

Ligule 64
Longevity 123
Lovastatin 37
Low-fluence 23

MSU1093 110
Maize 27
MAP kinase 78
Meristems 69, 93
Mevalonate 37
Mevinolin 39
Monocarpic 123
Monopteros 95

Norflurazon 59

Oligoglucosides 8
Oligosaccharides 8
Ovaries 106

P. savastanoi 46
Parenchyma
Pathogenesis-related proteins 16
Pattern formation 93
Pea 76
Periwinkle 100
Pertussis toxin 23
Petals 106
Petunia hybrida 87, 106
Phloem sap 17
Phosphorylation 24
Photoaffinity labeling 3
Photomor-phogenesis 21
Phyllotaxis 97
Phytoalexin 7
Phytochrome 57
Pinhead 70
Pistil 105
Pistillata 118
Polarity 93
Pollen 107
Polycarbonate barriers 102
Positional information 99
Proplastids 57
Protein kinase 131

R1 29

Receptor 11
Redifferentiation 100
Ribonuclease 107
Ribosomal proteins 78
RNAs 78
Root 95
Rootless 70

S-locus 105, 136
Salicylic acid 15
Seedling 93
Self-incompatibility 105
Senescence 51, 123
Sepals 117
SHAM 16
Shootless 70
Signal transduction 2, 7, 21, 45 54, 58, 131
Spadix 15
Styles 106
Suspensor 94
Superman 118
Systemically acquired resistance 16

TMK1 131
Tassel 64
Thermogenesis 15
Thidiazuron 39
Tissue culture 45
Tomato 52
trp1-1 mutant 80
Tryptophan monooxygenase 45
Tunica 69
Tyrosine kinases 131

Vascular differentiation 47
Viviparous 27
Voodoo lily 16
Vp1 29

Xylem 47

Zeatin 41
Zygomorphy 89